CW00467647

ICE Conditions of Contract for Minor Works

A user's guide and commentary

The Institution of Civil Engineers'

Conditions of Contract for Minor Works

A user's guide and commentary

Guy Cottam and Geoffrey Hawker

Thomas Telford, 1992

Published by Thomas Telford Services Ltd, Thomas Telford House,
1 Heron Quay, London E14 4JD

First published 1992

British Library Cataloguing in Publication Data
Cottam, G.
ICE conditions of contract for minor works: a
user's guide and commentary.
I. Title II. Hawker, G,
624.1

ISBN: 0 7277 1649 2

Typeset in Great Britain by MHL Typesetting Ltd, Coventry.

Printed and bound in Great Britain by Redwood Press Ltd, Melksham, Wiltshire.

FOREWORD

The ICE Conditions of Contract for Minor Works have now been in use for several years and it is timely that a user's guide and commentary is produced which draws upon experience to date and interprets the contract in the light of the provisions of the recently issued ICE Conditions of Contract sixth edition. When the decision was made to proceed with the preparation of conditions of contract for minor works there was a widespread feeling in the construction industry that the ICE Conditions of Contract fifth edition were not particularly suitable for certain classes of work of a minor or routine nature. In the event however the minor works conditions have not so far been entensively used and clients have continued to use the standard forms to which they are accustomed. This is a pity since there is much to commend the ICE Conditions of Contract for Minor Works in terms of simplicity and appropriateness for the types of work targeted. It is to be hoped that this book will have the effect of drawing the attention of potential users to the advantages of the conditions and therefore encourage their use.

The authors are widely experienced and respected engineers and arbitrators. Guy Cottam was formerly a director of an international firm of civil engineering contractors before becoming a consulting engineer, while Geoffrey Hawker is an engineer and barrister, and currently Head of Chambers. Their careful and detailed analysis and commentary provides an excellent introduction to the conditions as well as a work of reference which can be turned to from time to time. It will also be valuable to students.

It is a pleasure to commend this useful volume and its authors to the reader.

Stuart Mustow
Chairman, Conditions of Contract Standing Joint Committee
Vice President, Institution of Civil Engineers

CONTENTS

Part 1

INTRODUCTION

Chapter 1

INTRODUCTION

This book is intended to be a guide to all those involved in contracts under the Institution of Civil Engineers (ICE) Conditions of Contract for Minor Works — consulting engineers, contractors, sub-contractors and quantity surveyors. The book has been divided into three parts. Part I is an introduction to explain the background and reasons for the production of the Contract. Part II is a commentary on the clauses and is intended not as a legal treatise but as a guide to their interpretation. We consider that for minor works contracts, where the risk to the parties is small, the types of dispute that arise will not involve very large sums and the likelihood of going to arbitration is consequently small. It is important therefore to understand the reasons for the provisions rather than being inundated with a lot of references to cases which may or may not have direct bearing upon the interpretation of the particular clauses.

It is only a convention that cases resolved under one contract should be used for the interpretation of another and we feel that guidance on applying the clauses will be of more benefit to practitioners than careful argument on precise legal meaning. We have, however, provided at the end of each chapter, where appropriate, legal comments for those with a special interest in such matters. We have also given a comparison in each chapter between the provisions of the minor works form and those of the ICE Conditions of Contract (fifth edition) together with further comment where the provisions of the sixth edition differ from those of the fifth. The aim of these paragraphs is to guide those who are undecided as to whether to use the minor works forms or the ICE fifth or sixth (from hereon referred to as ICE 5 and 6 respectively) by comparing the relative risks to Employer and Contractor between the two forms of contract.

Part III is concerned with the application of the Contract. It consists of discussion on problems that arise and gives guidance as to how they can best be avoided. There are also two associated documents which will usually be used on contracts utilizing the minor works form. They are a standard method of measurement and a form of sub-contract agreement. None of the currently available standard documents is compatible with the requirements of the minor works form so we have included suggested amendments to ensure that at least one available standard document in both these categories can be modified for use with it.

The Civil Engineering Standard Method of Measurement, second edition (CESMM 2) has been designed for use with the ICE Conditions of Contract. There are certain matters in which the minor works form and ICE 5 are not compatible and as a result certain amendments will have to be made to

CESMM 2 if it is to be used with the minor works form. Details of these amendments and the reasons for them are to be found in Chapter 16. Also in Chapter 16 will be found methods of incorporating daywork schedules into the Contract whether or not Bills of Quantities are used.

The standard Federation of Civil Engineering Contractors Form of Sub-Contract (commonly known as the 'blue form') for use in conjunction with the ICE Conditions of Contract was also drafted for use with ICE 5 and certain amendments need to be made to it in order to make it compatible with the minor works form.

Discussions have been held with the FCEC and certain modifications agreed with them so that the blue form is compatible for use with the minor works form. These amendments are detailed in Chapter 17.

Our basic aim has been to try to avoid disputes arising under contracts using the minor works form by giving the participants greater understanding of the intention and meaning of the clauses, and by providing guidance on the production of the contract documents associated with it so that misunderstandings will not arise.

Chapter 2

BACKGROUND

Construction is a very ancient pursuit. Early humans, and indeed animals, have throughout time built themselves shelter from the weather. In early times construction was a DIY operation but as knowledge widened so the complexity of the structures we were able to construct increased and with it building became a speciality pursuit.

Whenever work is commissioned there necessarily must be an agreement as to what it is that the builder is required to do and what he will be getting for doing it. The need for conditions of contract has therefore been with us for a very long time. Standard conditions of contract are, however, a comparatively recent innovation. Prior to the introduction of standard conditions architects, engineers and clients wrote their own, which as may be expected tended to safeguard the interests of the writer of the contract by passing all the risks to the other party. This was not very satisfactory, and the courts started to find ways of circumnavigating their more outrageous requirements.

However, it is not only the conditions of contract themselves that have to be decided but also the procedures for inviting tenders and dealing with them once they have arrived. To promote understanding and trust it is necessary that these procedures be understood by the contracting parties. Many different methods of inviting tenders have been tried over the years, some of them pretty bizarre. In Roman times the architect would announce the date and time when he would receive tenders and on the appointed day would sit behind a desk upon which there was a lighted candle. He would then read the entire contract documents, the conditions and the specification and would accept offers until the candle went out. The mad scramble at the end whilst the candle flickered must have been an engaging sight. This system had the obvious benefit of ensuring brevity in the written word. Perhaps our current trend towards standard specifications would benefit from the re-introduction of this system. The first steps towards standard conditions were made in the building industry in the 1920s and civil engineering was to follow a little later.

The father of the ICE Conditions of Contract was Mr E. J. Rimmer, a civil engineer and a barrister at law who produced a paper which was published in the *Proceedings of the Institution of Civil Engineers* in February 1939 entitled 'The Conditions of Engineering Contracts'. This paper led to the publication of the first edition of the ICE Conditions of Contract in 1945.

Three points from Mr Rimmer's paper bear repetition for they are as true today as they were when they were written some fifty years ago.

> . . . it must be emphasized that the nearer the contract drawings can be prepared

> to represent the work as eventually carried out, the less likelihood there is of disputes arising under the contract.
>
> . . . Upon the clarity with which Engineers are able to express their requirements may depend much of the successful outcome of an engineering contract, and time spent on making a specification as precise and unambiguous as possible is time very well spent in the interests of the client.
>
> . . . One important word of warning must, however, be given. The basic principles upon which standard conditions should be framed must first be decided by the parties concerned. They may, amongst persons of different views, be determined on the basis of compromise, but if the conditions are to be clear and unambiguous there is no room for compromise on the manner and phraseology in which these principles are to be stated. This must lead, in the Author's opinion, to the overthrow of some clauses and phraseology now taken to be standard practice, but which, in fact, are confusing and ambiguous.

Whether such advice was heeded in drafting the first edition of the ICE Conditions of Contract may be doubted.

The first edition of the ICE Conditions of Contract was a joint document agreed between the Institution of Civil Engineers and the Federation of Civil Engineering Contractors (FCEC). All subsequent editions have in addition been agreed and sponsored by the Association of Consulting Engineers (ACE). These three bodies then formed a standing joint committee, know as the Conditions of Contract Standing Joint Committee (CCSJC) to monitor and keep under review the working of the ICE conditions.

The first major revision to the ICE Conditions of Contract was made in 1973 when the fifth edition was published. Up to that time the intervening editions were really only amendments. The fifth edition itself was something of a compromise. A draft of an entirely new contract had been prepared but was finally rejected in 1971 and the three sponsoring bodies then decided to undertake a major revision of the fourth edition instead.

This review resulted in the fifth edition (ICE 5) as published in June 1973. It was only subject to minor amendment until 1988 when a major review was undertaken. This review resulted in the sixth edition (ICE 6) published in 1991. This latter edition, although extensively amending the fourth, has not fundamentally altered the balance of risk between the parties, and the amendments generally reflect the changes that have occurred in the industry in the twenty years that elapsed between the consideration of the two editions.

When published, ICE 5 received a mixed welcome. Many of those representing employers and some lawyers took the view that the new edition was a 'Contractors' Charter' in that it provided additional scope for the reimbursement of extra costs. However, hand in hand with these provisions was set up a very much stricter routine for the giving of notices and the requirement for keeping records. Many of the doubts voiced at the time have not led to major problems and the record of the conditions as far as disputes are concerned is good, judging from the very small number of contracts under them that have been recorded in the law reports.

However, where a contract does permit grounds for additional payment it is necessary that the procedures for recording facts relevant to the additional payment are strictly adhered to. This in itself is a costly operation and a realization developed in the early 1980s that ICE 5 was somewhat heavy-handed when applied to small straightforward works. In 1983 the ICE's Advisory Committee on Contract Administration and Law (CCAL) — the committee then responsible for reviewing legal matters — decided that there was a need, within an 'ICE family of contracts', for additional standard contracts for minor works, design and build contracts and for maintenance works (the last two being areas of increasing importance to the civil engineer due to changing market requirements).

The need for a minor works contract was prompted by a desire to reduce the amount of paperwork that was being created on most major contracts and it was the intention of the drafting committee, right from the start, to produce a simplified standard contract for minor works based upon the principles inherent in the ICE Conditions of Contract without materially changing the balance of risk between the parties.

The intention was primarily to reduce the complexity of administrative procedures for the kinds of project for which simplification was deemed to be acceptable. An early draft of the proposed minor works form was presented at a seminar held in London, the results of which gave rise to a number of alterations, in particular the retention of provision for unforeseen physical conditions and artificial obstructions. The revised draft was then submitted by CCAL to CCSJC, following which it underwent further modifications culminating in the presentation of a penultimate draft to a second London seminar. Finally, after further consideration of comments made at the seminar, together with some comments submitted in writing after it, the minor works contract as it now stands was published in 1988, with a launching conference on 20 April 1988.

Although the Institution of Civil Engineers had been the prime mover in the production of this new contract, the final draft was agreed by the CCSJC and published as a joint document. The Committee also pioneered two new procedures. First the contract was issued with Notes for guidance on how it should be completed. These Notes for guidance were not intended to explain the clauses but merely to assist those who have to complete variable parts of the contract on how this ought to be done. The aim was to try to avoid at source mistakes that could cause difficulty later on. Experience had shown that quite a large number of people were incapable of understanding and completing correctly the Appendix to the Form of Tender for the ICE 5 and it was thought that guidance notes could have avoided a lot of the mistakes that had been, and were currently being made.

The second innovation was the introduction of conciliation as a stage in dispute resolution. Since the Engineer under the minor works form is to be in day-to-day control of the Works and will in many instances be the person instructing the Contractor it was considered that a formal reference back to him would serve little purpose. Many disputes, it was thought, that would arise would probably concern decisions that had already been taken by the Engineer and would have been carefully argued before they had become disputes. Thus formal reference back to the Engineer of a dispute as a condition precedent to arbitration (as in Clause 66(1) of ICE 5) was less likely to avoid the need for eventual arbitration and might degenerate to the point where such reference might become no more than a delaying tactic.

On the other hand, it was felt strongly that there should be some means of facilitation compromise, and thus avoiding the need for arbitration. It was therefore decided to eliminate the Clause 66 reference to the Engineer and put in its place a voluntary optional reference to a conciliator. This was to be the first standard contract with a non-binding stage written in to the dispute procedure.

A new ICE Conciliation Procedure was published with the minor works contract and, although designed initially for use with that contract, it can, by agreement, be used in conjunction with any form of contract. It is now incorporated into ICE 6 as an additional optional procedure.

The opportunity was also taken to write the contract in everyday English instead of the rather stilted and legalistic form displayed in ICE 5. Gone are most of the 'save', 'in respect thereof', 'pursuant to', 'notwithstanding' and the very long sentences which were to be found in most other standard contracts. A conscious effort was made to keep the need for cross-referencing between clauses to a minimum.

A contrast can be discerned by those who are interested in the language of the contract between Section 10 (Insurances) and the remainder of the contract. The reason is that Section 10 was imported almost totally from the ICE 5 so that the insurance provisions of both contracts would be obviously the same and to ensure that the insurance policies required for both types of contract would be interchangeable. It has, however, incorporated into the minor works form a noticeable contrast between two different styles of drafting.

An attempt was also made to rationalize the layout of the various requirements and collect related matters under sub-headings. Thus, for instance, all the powers of the Engineer are to be found in one clause, the obligations of the Contractor in another and matters of programming in yet a third. This is at variance with the ICE 5 where these powers and obligations are scattered throughout the various clauses in the document.

The contract, as published, if completed correctly, will contain all information necessary to define fully the scope of the contract and the documents that go to make it up. At the time of writing the contract has enjoyed wide penetration into the field of small works contracts and very few disputes have been reported.

The Conciliation Procedure has been adopted in a number of potential disputes and is rapidly gaining a reputation as a speedy and cost-efficient way of resolving disputes.

Chapter 3

LAYOUT OF THE MINOR WORKS FORM

Introduction

The Institution of Civil Engineers Conditions of Contract, Agreement and Contract Schedule for use in connection with Minor Works of Civil Engineering Construction First edition (January 1988) is the full title for the minor works contract. The Contract is an agreed document between the three sponsoring bodies for the ICE Conditions of Contract, which are the Institution of Civil Engineers, the Association of Consulting Engineers and the Federation of Civil Engineering Contractors. The document is published together with Notes for guidance and the ICE Conciliation Procedure 1988. The ICE Arbitration Procedure 1983 is also referred to in the Conditions of Contract but has to be obtained separately.

The contract comprises

(i) the Agreement
(ii) the Contract Schedule
(iii) the Conditions of Contract and
(iv) an Appendix to the Conditions of Contract

The minor works form does not contain a separate form of tender, but if one is thought desirable a sample form is given in Appendix E of this book, together with a suitable sample form of acceptance.

Pre-contract considerations

Intended use

The allocation of risk under the minor works form is broadly in line with the Institution of Civil Engineers' Conditions of Contract (fifth edition) (June 1973) as revised in January 1979 and reprinted in January 1986, commonly known as the ICE Conditions of Contract, fifth edition, and referred to in this book as ICE 5.

The main difference between the two sets of conditions is that the detailed administrative procedures and requirements for notices of ICE 5 have been removed in the minor works form. It follows that the minor works form is unsuitable for contracts where there could be major departures from either the design or methods of construction as envisaged at the time of tender. The form should not, therefore, be used when there are uncertainties concerning the Works, or where the Site may contain difficult ground conditions.

The contract is intended for works of a simple nature, for which the design is complete at the tender stage. Notes for guidance (NG) outline the intended use as follows

NG1. This form of contract is intended for use on contracts where:-

(a) the potential risks involved in the Works for both the Employer and the Contractor are adjudged to be small;
(b) the period for completion of the contract does not exceed six months except where the method of payment is on either a daywork or a cost-plus-fee basis;
(c) the Works are of a simple and straightforward nature;
(d) the Contractor has no responsibility for the design of the permanent works other than possibly design of a specialist nature (see Note 6);
(e) the contract value does not exceed £100 000;
(f) the design of the Works, save for any design work for which the Contractor is made responsible (see Note 6), is complete in all essentials before tenders are invited;
(g) nominated sub-contractors are not employed (but see Note 7).

The fundamental consideration of these seven points is that the potential risks involved in the works for both parties should be small, as given in Note 1(a). All the other six items are guidelines for potential risk assessment.[1]

The risks that should be considered when deciding whether or not to use the minor works form are as follows:

(i) For the Employer
Design
Is it complex?
It is likely to be subject to variation?
Is it complete in all major respects?
Does it consist of work of a specialist nature?
Is it clear in all its specification details?
Are any of the materials on long delivery, or likely to be difficult to purchase?

Contract documents
Have the quantities been able to be taken off accurately?
Are they likely to change?

Ground conditions
Is there any doubt about the conditions likely to be encountered?
Are there likely to be any artificial obstructions, such as previous foundations or services?

The Site
Is it available?
Is there likely to be any difficulty with access?
Could the operations have a serious effect upon any concurrent use of the site by the Employer?

---

1. However, there is in principle no reason to doubt that the minor works form could be used with success for large, but simple projects, or straightforward projects spread over an extended period (in which case some form of price variation may need to be added), provided that the other criteria are satisfied. There have, in fact, been several instances where such projects have been carried out satisfactorily using the minor works form.

(ii) For the Contractor

Resources

Is there likely to be any difficulty in obtaining labour, materials or plant?
Does the work involve any difficult or risky operations?
Are complex temporary works required?

Time

Is there likely to be any difficulty in completing in the proposed time?
Are the required Liquidated Damages a large amount relative to the contract value?

(iii) For both Contractor and Employer

Weather

Is the work likely to be seriously affected by adverse weather?

Third parties

Are properties or landowners adjacent to the site likely to be adversely affected by the construction of the works?
Is there likely to be opposition, or other problems of a political nature?
Is there likely to be excessive trouble from trespassers?
Could there be any difficulty in obtaining sufficient insurance cover to meet any potential risk?

Safety

Do the Works involve work of a particularly hazardous nature?
Do the Works involve dealing with materials of a hazardous nature?

This list in not comprehensive, and each project will have its own special risks to consider. It does, however, give examples of the sort of questions that the Engineer should satisfy himself about when deciding which form of contract to propose for a particular project.

NG1(b) the period for completion of the Contract does not exceed six months except where the method of payment is on either a daywork or a cost-plus-fee basis;

The period for completion is restricted to six months, because the longer the contract period, the greater the risk arising for the Employer.

Firstly, there is no frustration clause in the Contract, and the Employer's only remedy in the event of gross breach by the Contractor is to terminate at the completion date. Secondly, it is anticipated that the Contract will be let on a fixed price basis, and consequently a limited period is necessary to safeguard the interests of both parties. And thirdly, there is an increased chance of the financial position of the Contractor deteriorating over a longer period.

However, this time limitation does not apply to contracts let on a daywork or cost-plus-fee basis, because such contracts usually do not specify a finite amount of work, and can be terminated at the convenience of the Employer. The restriction therefore is unnecessary in such cases.

NG1(c) the Works are of a simple and straightforward nature;

The Works should be of a simple and straightforward nature because the risks involved for both parties clearly increase with increased complexity in either design or construction or both. Where the Works are complex, the provisions for Contractor's supervision and checking by the Engineer need to be more exacting than are required by this Contract. Assessment of what is complex is subjective, for what is complex to a general contractor may be simple and straightforward to the specialist, and when assessing what is simple and straightforward, reference should be made to the proposed list of contractors.

Where the Works are complex consideration should be given to using the full ICE 5 (or ICE 6), even where the project is otherwise of a minor works nature.

NG1(d) the Contractor has no responsibility for the design of the permanent works other than possibly design of a specialist nature (see Note 6);

The Contract is unsuitable for projects where the Contractor is to be responsible for the design. However, if the Contractor is required to take responsibility for the design of work of a specialist nature, then full details must be included in the Contract, defining exactly what the Contractor's responsibility is to be in respect of such work.

The work may be designed either by a specialist contractor or a specialist sub-contractor but if, since there is no provision within the Contract for the use of a nominated sub-contractor, a particular specialist is required for any specific items of work then that specialist must be named in the contract documents, in order that the Contractor may negotiate appropriate sub-contract terms prior to his tender.

Aspects of contractor design are covered for Note 6.

NG6. If the Contractor is required to be responsible for design work of a specialist nature, which would normally be undertaken by a specialist sub-contractor or supplier (such as structural steel, mechanical equipment or an electrical or plumbing installation) full details must be given either in the Specification or in the Appendix to the Conditions of Contract or on the Drawings indicating precisely the Contractor's responsibility in respect of such work.

The word to note is 'precisely'. Care must be taken when drafting this requirement that the duty imposed on the Contractor is not one of 'fitness for purpose', because it could render the Contractor's insurance policy void, but only one for the exercise of 'due skill and care'.

NG1(e) the contract value does not exceed £100 000;

This is a broad guideline, but there is no need to adhere strictly to it if the other criteria are fully met. There is, in any case, no direct relationship between risk and the value of the contract. It may be that contracts of a very much smaller sum may contain very high potential risk, and therefore should not be let on this form, but on ICE 5 (or ICE 6) instead.

NG1(f) the design of the Works, save for any design work for which the Contractor is made responsible (see Note 6), is complete in all essentials before tenders are invited;

The greatest potential risk from the Employer's side arises from incomplete design, which may lead to both extensive (and expensive) variation and delay to the Contractor whilst the design details are prepared and issued to the

Contractor. However, this guideline can be relaxed if the mode of payment is to be dayworks or cost plus a fee. The risks then for the Contractor are greatly reduced. The Employer has to balance the gain from an earlier start to construction against the potential risk of increased cost due to the absence of a firm price.

NG1(g) nominated sub-contractors are not employed (but see Note 7).

Since the intention is that the minor works form should only be used for contracts of short duration and of low risk, there is neither time required for the Contractor to negotiate with a nominated sub-contractor nor any logical reason to permit the use of nominated sub-contractors. Where nominated sub-contractors might be used, for instance for specialist works, these works must, therefore, be either designed fully by the Engineer so that the Contractor can obtain prices for them during the tender period, or included in the Contract for the Contractor to design and construct.

When the Contractor is to undertake the design of any part of the work outline details of the Engineer's design requirements must be included in either the Specification or the Tender Drawings. The details must be sufficient to enable the Contractor or a specialist sub-contractor to produce their design details from them. If the Engineer at this stage is uncertain what is required, then he would be better advised to contact a specialist supplier himself, agree the design details and name that supplier in the contract documents. This procedure is covered by Note 7.

NG7. The Engineer may in respect of any work that is to be sub-let or material purchased in connection with the Contract, list in the Specification the names of approved sub-contractors or approved suppliers of material. Nothing however should prevent the Contractor carrying out such work himself if he so chooses, or from using other sub-contractors or suppliers of his own choice, provided their workmanship or product is satisfactory and equal to that from an approved sub-contractor or supplier.

If the Engineer requires work to be sub-let to a specified sub-contractor, then the approved list of sub-contractors should be given in the Specification. This will enable the Contractor to negotiate with the specialist supplier prior to submitting his tender. However, where possible, the Contractor should always be free either to carry out the work himself or let the work to a sub-contractor of his own choice, provided that the Engineer is satisfied that the proposed sub-contractor can produce workmanship that is up to the required standard.

The intention is that the Employer should be able to take advantage of the Contractor's buying expertise. Without good reason, therefore, the source of such goods or materials should not be limited. If, however, one specialist supplier is required by the Engineer and agreed to by the Employer, then it would be permissible to name one sub-contractor, whom the Contractor would then be obliged to employ. The Contractor would then have to agree terms with that single sub-contractor during the tender period.

It must, however, be borne in mind that the Contractor may not be able to reach a satisfactory agreement with the single sub-contractor, and would, therefore, not wish to tender for the Works. It should be made clear in the Contract Documents that if the Contractor decides not to tender because of failure to reach agreement with such a single supplier this will not be held against him by the Employer when inviting future tenders.

Types of contract under the minor works form

The Contract is drafted for five alternative methods of payment and more than one method may be used on any one contract. These methods are

(i) lump sum,
(ii) measure and value using a priced Bill of Quantities,
(iii) valuation based on a schedule of rates (with an indication in the schedule of the approximate quantities of major items),
(iv) valuation based on a dayworks schedule,
(v) cost plus (the cost is to be specifically defined in the Contract and will exclude off-site overhead and profit).

Provision is made in Item 2 of the Appendix to the Conditions of Contract for the choice of mode of payment which will be selected by the Engineer.

NG3. The method of payment for the Contract should be as stated in the Appendix to the conditions of Contract but if a Bill of Quantities is used the method of measurement used must also be indicated in the Appendix. If a Daywork Schedule other than the 'Schedule of Daywork carried out incidental to Contract Work' issued by the Federation of Civil Engineering Contractors is to be used the Schedule to be used must be clearly identified in the tender documents.

The normal method of payment will be either lump sum or measure and value. The lump sum will be particularly appropriate where the design is complete and the likelihood of any variation minimal. Interim payments may be made on the basis of an assessment of the percentage of the Works completed at the date of Contractor's monthly statement.

When a Bill of Quantities is used, it is clearly necessary to specify a method of measurement and this is detailed in Item 3 to the Appendix to the Conditions of Contract. The method of payment will be selected to suit the type of work involved.

The schedule of rates, daywork or cost-plus bases will be appropriate where the work is of the nature of an emergency or the extent of the Works cannot be defined at the time of entering into the Contract. Also, where the scope of work involved in the Contract is not defined the lack of any power to determine the Contract is consequently of little effect.

Each of these bases for payment requires different treatment in the Contract documents.

Lump sum

A lump sum contract does not require a Bill of Quantities. The Contractor will normally be expected to take off the quantities from the drawings himself and take responsibility for their accuracy. Such a contract requires the design to be complete in all respects to enable the lump sum to be estimated, and also because the Contractor will be entitled to receive all design details at the start of the Contract. Since there will be no rates in the Contract as a datum for pricing variations may prove expensive as the Engineer and the Contractor may find their value difficult to agree.

However, where the works can be designed in their entirety before tenders are invited the lump sum contract should show savings because there is then no need for either party to undertake a detailed measurement of the Works. Interim payments may be made on the basis of an assessment of the percentage of the Works completed at the date of the Contractor's monthly statement.

Measure and value

This is the usual form of payment. The Bill of Quantities will be drawn up by the Engineer in accordance with the rules of a standard method of measurement which is to be named in Item 3 of the Appendix to the Conditions of Contract.

3. Where a Bill of Quantities or a Schedule of Rates is provided the method of measurement used is

. . . .

Any suitable standard method of measurement may be used for compiling the Bills of Quantities, but the Civil Engineering Standard Method of Measurement (CESMM 2) will generally be the most appropriate[1]

The effect of this provision is that the Bills of Quantities are warranted by the Employer to have been compiled in accordance with the named method of measurement, and any errors or deviations from the method (unless such deviations are noted in Bills of Quantities), will require to be corrected by the Engineer and the Contractor will be paid on the basis of the corrected Bills.

It must be remembered that the method of measurement has to be read as a whole, and there are sometimes items to be included in the Bills of Quantities for work which the Contractor may not carry out (i.e. back shutters for concrete walls, top shutters, etc.). These items must be measured and the Contractor must have paid for them whether or not he actually carries out the operations. The principle is that if the Contractor, at the time of tender, considers that he will have no requirement for the item he will price it at zero.

Rates for items of work which are omitted from the Bills in error will have to be valued, not on the basis of the cost of the work to the Contractor, but on what is a reasonable market rate for the type of work described.

Valuation based upon a Schedule of Rates

This alternative should only be adopted when the actual quantities of work cannot be measured with any degree of accuracy. Such works as emergency repairs of embankments, or replacement of damaged road surfaces (where the extent of the damage cannot be established until the surfacing has been removed) are suitable for this method of evaluation. The descriptions in the Schedule of Rates must be drawn up in accordance with the standard method of measurement defined in Item 3 of the Appendix to the Conditions of Contract so that the Contractor knows what operations have to be included in the price for each item.

The Engineer has to give the approximate quantities of the work involved

1. CESMM 2 will need to be amended if used with the minor works form and these amendments are detailed in Chapter 16.

so that the Contractor can gauge the extent of the work to be carried out and the resources (human, material, mechanical and financial) that will be required. Any large deviation from these approximate quantities could lead to a claim for mis-statement or misrepresentation, and an increase in the rates.

Valuation on a Daywork Schedule

There are three alternative ways of incorporating daywork rates into the Contract. These are either by a contract schedule with items covering the labour, materials and plant likely to be used on the Contract and which the Contractor prices in his tender; or by specifying that the rates contained in the 'Schedules of Daywork carried out incidental to Contract Work' issued by the Federation of Civil Engineering Contractors current at the date of the tender shall be used; or permitting the Contractor to quote a percentage adjustment to the rates given in the Federation Daywork Schedule. An example of each these methods is given in Appendix E.

If a schedule for the Contract is to be drawn up, it is essential that it should clearly define what is to be included in each rate. Such items as productivity bonus, overtime rates, wastage of materials, cost of supervision, cost of general site transport and general site overheads all lead to problems if the manner in which they are to be dealt with is not clearly specified in the preambles to the schedule.

If the FCEC daywork schedule is to be used, two approaches are possible. The schedule is divided into three sections for labour, materials and plant. Each section has a different percentage addition applied to the rates or prices in it to cover the Contractor's overheads and profits. The FCEC publishes suggested percentage additions. These additions can either be specified to be applicable in the contract documents or the Contractor can be asked to quote separately the percentage additions that he requires.

Where the Daywork Schedule is required to price only incidentals then it is probably appropriate to specify that the FCEC percentages shall be applicable. However, if the entire contract is to be valued on the basis of daywork rates then, in order to obtain some element of competition in the tendering process, the Engineer should assess the likely expenditure on the proposed work for labour, materials and plant as lump sums and invite the Contractor to add his own percentage additions to these lump sums, the total of the lump sums and the percentage additions then being the total of the Contractor's offer. The FCEC daywork schedule and how it should be used are explained in Chapter 16.

Cost plus fee

If this method is adopted it is necessary to define clearly the basis for the assessment of 'cost' and what the Contractor is required to cover in his 'fee'. There are several published contracts which will assist in defining cost and the 'Joint Contracts Tribunal (JCT) Fixed Fee Form of Prime Cost Contract' is a useful source for a definition of cost. The fee can be a fixed lump sum, a percentage of the eventual cost, or a variable amount calculated on the basis of some form of agreed target cost. However, since the work involved is likely to be low in value the first two alternatives should prove adequate for use on most occasions with this contract.

Inflation

The Contract should normally be let on a fixed price basis.

NG4. In view of their short duration all contracts should normally be let on a fixed price basis with no provision for price fluctuation. The letting of contracts on a daywork or cost-plus-fee basis is however not precluded (see Note 1(b)).

This of course means that the rates are not subject to adjustment for changes in costs due to inflation or other reasons. The fixed price basis must not be confused with a lump sum contract, which is not subject to remeasurement for changed quantities of work.

Other pre-contract considerations

There are a number of other matters which have to be considered either before selecting the minor works form as the contract conditions to use or before beginning to draft the contract documents prior to tender. These include

NG8. It is intended that acceptance should follow within 2 months of the date for submission of tenders.

NG9. Access as necessary to the Site should be available at the starting date under Clause 4.1.

The Engineer must check that sufficient time is allowed for the submission of tenders and tender appraisal, and that the necessary Site and access to it will be available to enable the Works to start on the starting date.

In this Contract, because it is anticipated that there will be a short period for completion, it is the responsibility of the Employer to provide the necessary access to, as well as possession of, the Site. The Engineer should ensure that there is sufficient access for any heavy plant that may be used.

The Engineer has to give consideration to these matters before deciding that the minor works form is suitable for the Works. He should also consider and, if necessary, discuss with the Employer the documents to be listed in the Contract Schedule, and the items to be entered in the Appendix to the Conditions of Contract.

Contract documentation

The Agreement

The Form of Agreement is given on page 1 of the Contract. The completion of the Agreement is optional. A Contract consisting of the Contractor's tender and written acceptance thereof by the Employer would be equally binding. However, completion of the Agreement is to be recommended, since Article 3 defines exactly the documents forming part of the Contract.

The Contract may be formed initially by acceptance by the Employer, by letter to the Contractor followed by an Agreement which is signed later. The date affixed to such subsequent Agreement is of little significance, since the effect of the Agreement would be back-dated to the commencement of the Contract [*Trollope and Colls Ltd* v *Atomic Power Constructions Ltd* (1962)].

The Agreement only provides for a simple contract. There is no provision for the Contract to be executed under seal, or as a deed. The difference between a simple contract and one under seal lies mainly in the effect of the Limitations Act 1980, the provisions of which limit the time for bringing an action for breach of contract from the date of the breach in the case of a simple contract to six years, and twelve years in the case of a contract under seal.

Since the Works will normally be of a simple and straightforward nature, use of a simple contract should suffice. If, however, for some reason a contract under seal is considered expedient, it will be necessary to add to the Agreement express words making it clear that the Agreement is executed as a deed, since the statute affecting the execution of deeds has recently been changed.[1]

Contract Schedule

The Contract Schedule lists all documents that are intended to form the Contract. The Schedule does not contain provision for amendment or addition to the Conditions of Contract, which should be avoided.

NG5. The procedures for letting and administering a minor works contract are intended to be as simple as possible in line with the low risk involved. The Contract should be fully defined in the documents listed in the Contract Schedule. There is no provision for amendment or addition to the Conditions of Contract and this should be avoided.

The items in the Contract Schedule are

Agreement. The Agreement need not necessarily be signed. A letter of acceptance is perfectly adequate to form a valid Contract.

Contractor's Tender. It should be noted that any general or printed conditions attached or referred to in the tender are not incorporated into the Contract unless their incorporation is expressly agreed in writing. This does not prevent the Contractor from making conditions of his own in a letter accompanying his tender, when, since these are a part of his tender, they will be included as part of the Contract.

Conditions of Contract. No additions, deletions or substitutions are required, and they should not normally be allowed.

Appendix to the Conditions of Contract. (See below).

Drawings. Since the Contract is to be fully designed before going to contract, all the working drawings should be listed in the Contract Schedule.

Specification. The reference should include the number of pages and any identifying title to the documents forming the Specification.

Priced Bill of Quantities/Schedule of Rates/Daywork Schedule. These items should be identified if included in the Contract, but their inclusion will be dependent upon the form of payment chosen.

---

1. By section 1 of the Law of Property (Miscellaneous Provisions) Act 1989 for individuals and section 130 of the Companies Act 1989 for companies. Reference should now therefore be made to 'deeds' rather than 'contracts under seal'.

Following letters. Any correspondence that has passed between the parties either prior or subsequent to the Contractor's tender which have a bearing upon his price may be included here. If so, care should be taken to ensure that there is no ambiguity between the various letters that are incorporated. If there are a number of such letters it is generally easier, clearer, and less likely that a mistake will be made if the essential points contained in the correspondence are summarized by the Engineer in a letter to the Contractor prior to the acceptance of the tender. Only this letter and the Contractor's written acceptance of it should then be listed in the Contract letters.

The intention is that all documents forming part of the Contract should be listed in the Contract Schedule.

NG2. The Contract Schedule should list all documents that will form part of the Contract. It is particularly important to ensure that the Appendix to the Conditions of Contract is prepared before tenders are invited (and the Appendix must be included with the documents supplied to prospective tenderers). Notes on the completion of the Appendix are included in these guidance notes (see Note 13).

This Note infers that it may possibly be desirable to include the Conditions of Contract in the tender documents by reference and not send out a full copy of the conditions. The minimum requirement would be a copy of the Contract Schedule and a copy of the Appendix to the Conditions of Contract. The schedule should be completed to include all tender documents intended for inclusion in the Contract, and the Appendix should be completed as far as possible with the information available at that time.

It will be seen that the Contract Schedule includes letters that have passed between the parties and the Engineer during the post-tender and pre-contract stage. It therefore will often be necessary to re-issue the Contract Schedule with the acceptance letter so as to incorporate the necessary correspondence.

Conditions of Contract

The conditions are complete in themselves and, unlike some other standard contracts, there is no provision for spaces to be completed, or alternatives to be deleted. Amendments to the standard conditions should be avoided, unless they are absolutely necessary, as it is very easy to overlook possible knock-on effects of any amendment on other unamended provisions in the Contract.

Appendix to the Conditions of Contract

There are 16 matters covered in the Appendix and Note 13 (which is discussed in detail below with the items to which they refer) assists the Engineer in completing the Appendix.

1. Short description of work to be carried out under the Contract. This description defines broadly the extent of the work included in the Contract. It therefore limits the scope of the variations that the Engineer is permitted to instruct, and the work which the Contractor is obliged to carry out. Care should be taken to ensure that it defines precisely what is required.

2. The payment conditions. There are a number of different methods of payment permitted under the Contract and these are defined by deleting the alternatives not applicable in the Appendix. Any number of these alternatives can be used in any one contract.

3. Method of measurement. Where a Bill of Quantities is used then the method of measurement must be defined. This will include the date and edition as well as the title. It is possible that more than one method of measurement may be used, in which case each of the methods of measurement used must be named in the Appendix. At the beginning of each bill it should be stated under which method of measurement it has be compiled.[1]

4. Name of the Engineer.

NG13(1). *Clause 2.1 (Name of the Engineer).* It is the intention that the name of the Engineer who will personally be responsible for the Works should be stated.

It is essential for the smooth running of the project and the administration of the Contract that the Contractor should know precisely with whom he has to deal, particularly as the Engineer may have no continuing representation on Site. The naming of a firm or company as Engineer is thus less than satisfactory, since many persons within that organization will know little or nothing of the particular project in question. It is for this reason that the person directly responsible for the project should be named as Engineer.

5. Starting date (if known). The starting date should normally be within 28 days of the date of acceptance of the tender. In any event, it is desirable that the starting date should be known at the time of tender.

6 and 7. Periods for completion.

NG13(2). *Clause 4.2 (Period for completion).* If the Contract requires completion of parts of the Works by specified dates or within specified times such date or time and details of the work involved in each part must be entered in the Appendix to the Conditions of Contract.

If there is to be more than one date (or time) for the completion of the Works it is essential that each part carrying a different completion date is properly described, since any element of the Works which is not so described will be affected only by the 'whole Works' completion date. For this reason, it would be prudent to provide for either a single completion date for the whole of the Works or a series of separate completion dates for designated parts of the Works together with a final date for 'all parts of the Works not otherwise described'. Part completion dates with an overall 'whole Works' completion date could in some circumstances lead to doubt.

The effect of a stated completion date is that the Contractor is not only bound to complete the Works or the relevant part thereof by that date (unless the Engineer grants an extension of time) but is also *entitled* to the full contractual period so stipulated, whether or not he intends to use it all. It follows that the Engineer has no power to instruct the Contractor to complete any part of the Works before the relevant completion date unless provision to that effect is written into the Appendix to the Conditions of Contract, paragraph 7.

---

1. See also Chapter 16.

8 and 9. Liquidated damages.

NG13(3). *Clause 4.6 (Liquidated damages and limit of liquidated damages).* A genuine pre-estimate of the likely damage caused by any delay should be assessed and reduced to a daily or weekly rate. The limit of liquidated damages should not exceed 10% of the estimated final contract value and this should be taken into account when assessing the daily or weekly rate.

A genuine pre-estimate of the likely damage is essential if the damages figure is not to be challengable in court. The minimum figure should be the loss of interest on the capital sum, together with the running costs of the Engineer's site supervision. There may also be costs associated with claims by other direct contractors for delays caused to them, and in the case of commercial activities, loss of income as well.

For practical reasons the limit of liquidated damages should not exceed about 10% of the estimated final contract value. The assessment of a weekly rate should therefore take this sum into account. It is counter-productive to assess the damage figure so high that the limit is reached in a very short period, after which the incentive effect of liquidated damages as an aid to completion will be lost.

Smaller contractors may not be able to meet a substantial figure and it is unlikely that there would be any gain to the Employer in putting the Contractor into liquidation.

10. Defects Correction Period.

NG13(4). *Clause 5.1 (Defects Correction Period).* This should normally be 6 months and in no case should exceed 12 months.

The longer the Defects Correction Period the longer it is before the final account can be settled. With minor works, where the risks are expected to be small, it is in the interest of both parties that the final account should be settled whilst details are fresh in the minds of those who were involved. That is why the six months is recommended and, since the end of the Defects Correction Period only terminates the mutual obligation placed upon the Parties in relation to repairs (the Contractor's duty to do them, and the Employer's duty not to employ others for their execution) there should be little actual risk involved for the Employer.

11 and 12. Rate and limit of retention.

NG13(5). *Clause 7.3 (Rate of retention and limit of retention).* The rate of retention should normally be 5%. A limit of retention has to be inserted in the Appendix and this should normally be between the limits of 2% and 5% of the estimated final Contract value.

In the construction industry the normal range of Contractor's profit is between about 5% and 7% of the Contract value. A larger retention than the proposed 5% means that the Contractor will not be in profit (and could well face losses) until he receives the first tranche of the retention money. Its effect would be to increase the amount of money that the Contractor has to provide to complete the works, and on which the Employer will be charged interest in the Contractor's tendered rates. It may also present the Contractor with cash flow difficulties. None of these are beneficial to the Employer's interest.

Note that a figure has to be written into the Appendix for the limit of retention, which the Engineer has to calculate before sending the documents out for tender. This is done by multiplying the estimated Contract value by the selected percentage for the limit of retention.

13. Minimum amount of interim certificate.

NG13(6). *Clause 7.3 (Minimum amount of interim certificate).* It is recommended that the minimum amount of an interim certificate should be 10% of the estimated amount of the final Contract value rounded off upwards to the nearest 1000. This minimum only applies up to the date of practical completion of the whole of the Works.

The minimum amount of an interim certificate is only needed for administrative purposes to avoid the work involved in issuing certificates for small amounts. For short contracts and those employing the smaller contractors there is good argument for not having a minimum amount at all. The Contractor is in any case required to submit a monthly statement regardless of whether its value reaches the minimum limit (Clause 7.2).

The minimum limit is only effective up to the issuance of the certificate of practical completion. This is in line with ICE 5. The reason for this is because the Contractor is obliged to carry out uncompleted work during the Defects Correction Period and there is no reason why he should not be paid for it, whatever its value, when it is carried out; he should not have to wait until the final account is certified.

14. Bank name. This is the name of the bank whose lending rate will be used to calculate the interest due on overdue certificates. It is suggested that one of the big high street banks should be named.

15. Insurance of the Works.

NG13(7). *Clause 10.1 (Insurance of the Works).* This is at the option of the Employer. It must be borne in mind that Contractors frequently carry large excesses on their all-risks policies so that the Contractor then accepts the risk under Clause 10.1 in respect of any uninsured loss. When the insurance under Clause 10.1 is to be provided by the Employer the details of such insurance, including any excesses which the Contractor may be expected to carry, should be stated in the tender documents.

The Employer must decide whether it is he or the Contractor who is to take out all-risks insurance. If the Employer takes out such insurance, then full details of the policy must be provided in the tender documents, particularly those provisions concerning excesses or any other liability that the Contractor is obliged to carry. The policy must be in the joint names of the Contractor and the Employer. The Engineer must strike out either 'required' if the Employer is to provide the insurance, or 'not required' if the Contractor is to take out the insurance. The subject of insurance is discussed in Chapter 13.

16. Minimum amount of third party insurance.

NG13(8). *Clause 10.6 (Third party insurance).* A minimum cover of £500 000 for any one accident/unlimited number of accidents should normally be insisted upon. In certain locations where there is greater risk to adjacent properties a higher limit may be desirable.

An assessment should be made of possible third party claims and a realistic estimate made for damages that they would involve. Work, however small, on an airfield, for instance, could provide a hazard of many millions of pounds. Equally, contracts for underpinning, or those close to adjacent buildings could cause damage to the buildings and possible loss of earnings if the buildings had to be evacuated as a result of the Works. The suggested £0·5 million is a reasonable amount and most Contractors will carry more than that.

In filling in the Appendix only a sum of money is required for the value of insurance required. The 'any one accident/number of accidents unlimited' is a fixed requirement and remains unaltered.

Tendering procedure

There is no designated procedure for tendering published for use with the minor works contract. However, there is no reason why the procedures set forth in *Guidance on the Preparation, Submission and Consideration of Tenders for Civil Engineering Contracts* published by the ICE Conditions of Contract Standing Joint Committee should not be followed. Copies of this publication can be obtained from the Institution of Civil Engineers.

To enable Contracts to be let quickly after the submission of tenders, it is advisable that the contractors tendering should be pre-selected and that open tendering should not be used. A minimum period of four weeks should be allowed for tendering but consideration should be given to increasing this period if the tendering contractors have to negotiate with a sub-contractor named in the Specification.

As a rule, a starting date should be specified in Item 5 of the Appendix to the Conditions of Contract, and only in exceptional circumstances should this be left blank.

The instructions to tenderers, which accompany the tender documents, should draw the tenderers' attention to such matters as

(i) the insurance requirements in accordance with Clause 10
(ii) the name of the person (with a telephone number) dealing with queries
(iii) arrangements for inspecting and visiting the Site
(iv) the procedure to be adopted in presenting and submitting tenders
(v) the approximate date when the successful tenderer will be informed, and
(vi) a note informing tenderers that qualified tenders will not be considered.

Acceptance should take place as soon as possible after the submission of tenders to enable the Contractor to have as much time as possible for the purposes of pre-planning. It is the intention that acceptance should follow within 2 months of the submission of tenders. (In Notes for guidance see Note 8.)

Since the Works have to be fully designed before going out to tender, the Contractor should be in possession of all necessary information to enable the planning to start as soon as notification of acceptance of his tender has been received.

If the starting date is not given in the Contract the Engineer should, if possible, notify the Contractor of the starting date at the time of notification of acceptance of his tender. The starting date normally has to be within 28 days of acceptance of the tender (Clause 4.1). The Engineer should ensure that all the Site is available for the use of the Contractor from the starting date (Note 9).

With the notification the Engineer should request

(i) the name of the Contractor's agent (Clause 3.4),
(ii) the name of the Statutory Safety Supervisor,
(iii) the Contractor's programme as specified in Clause 4.3,
(iv) the name of Contractor's insurers and proof of the Contractor's insurances (cover note) (Clauses 10.6 and 10.7),
(v) notification of which works the Contractor intends to sub-let (Clause 8.2),
(vi) a convenient date for an inaugural meeting, and
(vii) the Contractor to sign the form of Agreement on page 1 of the Contract (if required) and return it to the Engineer.

He should also provide the Contractor with the following information:

(i) the name of the Resident Engineer (if any) and the powers delegated to him (Clause 2.2),
(ii) the names of other persons with delegated powers and the limits of such powers (Clause 2.2),
(iii) any facilities required for other Contractors and the dates from which they are required (Clause 3.9).

In the Notes for guidance Note 11 points out

> NG11. In respect of Clause 3.4 it has to be recognized that in a minor works contract the Contractor might have no full-time supervisor on site and the Contractor may ask for instructions to be delivered or sent elsewhere for the attention of his representative. In these circumstances the Contractor has to accept the fact that urgent instructions might in the interests of safety or for some other reason have to be given directly to the Contractor's operatives on site.

It may also be that the Engineer does not have full-time supervision on site either and it will be necessary for the Contractor's agent and the Resident Engineer (or the Engineer) to make arrangements to enable them to keep in regular contact for the purposes of inspection, testing and measurement. Such matters should be discussed and appropriate arrangements agreed at the inaugural meeting.

Finally, the Contract does not state the number of copies of the contract documents that are to be provided to the Contractor. It is suggested that three copies of the contract documents, and of any revised drawings or specifications should be supplied by the Engineer. This is particularly important for drawings since the Contractor is unlikely to have any suitable copying facilities on site.

Part 2

COMMENTARY

Chapter 4

DEFINITIONS

Introduction

The number of definitions has deliberately been kept small, partly as a result of drafting policy and partly because the minor works form itself is intended to be a simple document.

The definitions

1. *Works*

1.1. 'Works' means all work necessary for the completion of the Contract including any variations ordered by the Engineer.

There is no distinction made between permanent and temporary works, both being included in the definition of the Works. However, the only time this distinction is of any significance is in relation to design and this is covered by Clause 3.7.

It should be noted that the Engineer's power to order variations is confined to the purpose of the Contract and this definition of the Works does not entitle the Engineer to broaden the scope of the Contract by ordering variations which are clearly not within the scope of the Contract as originally envisaged.

2. *Contract*

1.2. 'Contract' means the Agreement if any together with these Conditions of Contract the Appendix and the other items listed in the Contract Schedule.

The Agreement is optional. The Contractor's tender and the Employer's letter of acceptance is all that is necessary to form a binding contract. For documents to be incorporated into the Contract they must to be listed in the Contract Schedule (the Contract Schedule is discussed in Chapter 3). Since there is no standard form of tender published with the minor works contract for the Contractor to complete, it is important that the Contractor details in his tender letter all the documents that he intends to include as part of his offer so that they do get incorporated into the Contract. If they are not so listed there is a danger that they may get overlooked or mislaid and may then be deemed not to form part of the Contract [*Davis Contractors* v *Fareham UDC* (1956)]. A sample form of tender, suitable for use with the minor works contract, is given in Appendix E.

Care should be taken when listing the documents for inclusion in the Contract Schedule that the additional documents are really intended to form part of the Contract and that the implications of including them are fully understood. For instance, the Contractor might submit a draft programme with his tender. If this is included in the Contract Schedule it will thereby become a contract document, and every detail of that draft programme will be legally binding on both parties. Thus the Contractor will have no discretion whether to follow that programme or not and if he did depart from it the Employer could claim compensation. Similarly, if the Engineer failed to provided information to enable the Contractor to follow the programme exactly, the Employer would be in breach of the Contract, and liable to the Contractor for any loss or delay arising from that breach.

Similarly, if the Contractor's method statement is included in the Contract Schedule (even if only by cross-reference) the Contractor will not only be entitled but obliged to use that method and none other [*Yorkshire Water Authority* v *Sir Alfred McAlpine & Son (Northern) Ltd* (1985)]. Such an inclusion can cut both ways, with unforeseen consequences for Contractor and Employer alike.

3. Cost

1.3. 'Cost' (except for 'cost plus fee' contracts, see Appendix) includes overhead costs whether on or off the Site of the Works but not profit.

This definition is required for the evaluation of additional payments due to the Contractor for delay and disruption in accordance with Clause 6.1. The overhead costs recoverable are site overheads and head office overheads, but profit is not included. The exception in the brackets is only applicable to a contract let on a cost-plus-fee basis in accordance with Item 2(e) of the Appendix to the Conditions of Contract.

3. Site

1.4. 'Site' means the lands and other places on under in or through which the Works are to be executed and any other lands or places provided by the Employer for the purpose of the Contract.

Since the definition of Works includes the temporary works any land provided by the Contractor upon which he either constructs parts of the permanent works or prefabricates his temporary works will come within the definition of the Site. This is important because it can enable the Contractor to claim in his monthly statement (Clause 7.2) the cost of material delivered to his yard or work done in it. The Engineer for his part will have a right of entry into the Contractor's yard for inspection purposes.

It is also necessary for the Employer to supply (or arrange access to) sufficient land to enable the Contractor to construct his reasonable temporary works (e.g. scaffolding to perimeter walls, piling to foundation walls, and supports to overhangs).

4. Excepted Risks

1.5. 'Excepted Risks' are riot war invasion act of foreign enemies hostilities (whether war be declared or not) civil war rebellion revolution insurrection or military or usurped power ionizing radiations or contamination by radioactivity from any nuclear fuel or from any nuclear waste from the combustion of nuclear fuel radioactive toxic explosive or other hazardous properties of any explosive nuclear assembly or nuclear component thereof pressure waves caused by aircraft or other aerial devices travelling at sonic or supersonic speeds or a cause due to the use of occupation by the Employer his agents servants or other contractors (not being employed by the Contractor) of any part of the Permanent Works or to fault defect or error or omission in the design of the Works (other than a design provided by the Contractor pursuant to his obligations under the Contract).

This definition is required in connection with the provisions for insurances, Section 10 of the Conditions of Contract. Insurance is discussed fully in Chapter 12.

In general, the Excepted Risks are those which the Contractor is not required to bear and which therefore fall upon the Employer. They can be roughly classified into outside events such as war, riot and rebellion; unusual risks such as explosions, nuclear radiation and sonic bangs; and matters which are peculiar to the Employer or the project, such as design faults or use or occupation of the Works by the Employer or his agents.

Comparison with ICE 5

The definition of Site is a wider definition than ICE 5 by reason of the wider definition of the Works. This could clearly cover any lands provided by the Contractor, which are not included in ICE 5.

The definition of Cost in both contracts has the same intention, that is, that both site and head office overheads are included in the definition, but profit is not. The minor works contract's definition is the more specific.

The Excepted Risks are defined in language identical to that used in Clause 20(3) of ICE 5. This accords with the principle that insurance policies taken out annually for contracts under ICE 5 should also be suitable for contracts under the minor works form. The fact that the definition of Excepted Risks appears in the definitions in the minor works form and not in the body of the indemnity clauses as in ICE 5 has no significance.

A large number of the definitions in ICE 5 have been omitted from this contract. Most of them are self-explanatory in any case. What has been omitted can be seen by reference to the clause comparison in Appendix B.

Comparison with ICE 6

Whilst there are minor differences between ICE 5 and ICE 6 there is nothing of great significance to the definitions common to both ICE 6 and the minor works form.

The Engineer can now specifically be a firm, the Completion Certificate is now called the Certificate of Substantial Completion, the Maintenance Period, as in the minor works form, is called the Defects Correction Period and the Construction Plant is renamed Contractor's Equipment. These name changes have been made to convey a clearer impression of their intent.

The definitions of Provisional Sum, Prime Cost Item, and Nominated Subcontractor have been moved from Clause 58 to the Definitions Clause 1. However, the one important change is that the Engineer is given the power to agree that lands outside the Site shall be designated as part of the Site. No such power is given to the Engineer in the minor works form.

The relationships between ICE 6 and the minor works form are therefore broadly as set out above.

Chapter 5

ENGINEER

Introduction

The role of the Engineer lies at the heart of the Contract. The Engineer carries out three functions:

(i) he normally designs the Works;
(ii) he is required to ensure that the Works as constructed comply with the Specification and Drawings; and
(iii) he issues certificates as the Contractor becomes entitled to them (including certificates for payment) and agrees or determines all payments due to the Contractor.

However, he does not have the quasi-arbitral role under the disputes procedure in Clause 66(1) of ICE 5, since this is replaced in the minor works form by a new Conciliation Procedure which is discussed fully in Chapter 14.

In all these roles the Engineer acts as Agent for the Employer. The Engineer is employed by the Employer, either under a Contract of Engagement in the case of an independent consulting engineer or a Contract of Employment if the Engineer is an employee of the Employer. His primary duty is therefore to the Employer.

However, the Contract (and the ACE Terms of Engagement, under which an independent consulting engineer is normally contracted) requires the Engineer to act fairly between the Contractor and Employer, particularly when certifying additional payments. Indeed for a successful outcome the Engineer *must* act fairly. The contract gives the Engineer very wide powers to instruct the Contractor on matters concerning the Works, and the only criterion for evaluation of the Contractor's entitlement to additional payment, which may result from these instructions, is what is 'fair and reasonable' in the opinion of the Engineer.

The legal doctrine of privity of contract usually means that a contract confers neither rights nor duties upon persons who are not themselves party to that contract. However, like most standard forms of contract in the construction industry, provision is made in the minor works form for the Contract itself to be modified from time to time (albeit in a limited and closely regulated manner) by a 'stranger' to the Contract, namely the Engineer. Put another way, the Engineer is not a party to the Contract but he is a 'creature' of it, in the sense that his position is created by the Contract and he, therefore, derives all his powers from it.

Thus any failure of the Engineer to act fairly will not entitle the Contractor to take action against the Engineer for breach of duty of care in tort [*Pacific Associates* v *Baxter* (1989)] but may put the Employer into breach of the

Contract with the Contractor, enabling the Contractor to proceed for damages against the Employer. If the Employer is thereby put to unnecessary expenditure as a consequence of the Engineer's acts the Employer may be able to recover such costs from the Engineer.

The Engineer, as designer, is responsible for the supply to the Contractor of all details necessary for the construction of the Works (Clause 3.6). Since it is a principle of this Contract that the design should be complete before being put out to tender (Note 1(f) in the Notes for guidance) there is no provision for the Contractor to give notice of any further details required. Failure to provided necessary details when they are required by the Contractor may, and usually will put the Employer into breach of the Contract and if the Contractor is held up through lack of detail he will be entitled to recover any damage suffered thereby.

The Engineer

Naming the Engineer

Who is the Engineer?

2.1. The Employer shall appoint and notify to the Contractor in writing a named individual to act as Engineer. If at any time the Engineer is unable to continue the duties required by the Contract the Employer shall forthwith appoint a replacement and shall so notify the Contractor in writing.

The intention of the Contract is that the Engineer is to be a named individual who will be in charge of the Works and with whom the Contractor can discuss any problems that arise during their construction. Item 4 of the Appendix to the Conditions of Contract requires the Engineer's name to be inserted into the Contract, and Note 13(1) confirms this intention.

NG13(1) *Clause 2.1 (Name of Engineer)*. It is the intention that the name of the Engineer who will personally be responsible for the Works should be stated.

It is neither the intention that the Engineer should be a mere figurehead, nor that a firm or partnership should be designated as Engineer. This is confirmed by the requirement that the Employer shall notify the Contractor in writing of the appointment of a replacement if the Engineer is unable to continue his duties.

Delegation of powers and the Resident Engineer

The Engineer is entitled to appoint a named Resident Engineer or other experienced person to watch and inspect the Works and to delegate any or all of his powers to that person (Clause 2.2). However, when any instruction is issued either by the Engineer or an authorized delegate, the Contractor is entitled to ask under which power in the Contract such instruction has been given (Clause 2.4).

Both the Employer and the Contractor undertake to be bound by and give effect to every decision or instruction of the Engineer subject to the matter being referred to arbitration (Clause 2.7).

2.2. The Engineer may appoint a named Resident Engineer and/or other suitably experienced person to watch and inspect the Works and the Engineer may delegate to such person in writing any of the powers of the Engineer herein provided that prior notice in writing is given to the Contractor.

The Engineer is given authority to delegate all of his powers if he so wishes. However, Note 10 suggests that he will not normally delegate certain of his powers.

NG10. In respect of Clause 2.2 it should be noted that in all normal circumstances the Engineer would not be expected to delegate his powers under Clauses 3.8, 4.4, 5.4, and 7.6.

These clauses are:

3.8 Certifying adverse physical conditions or artificial obstructions,
4.4 Awarding extensions of time for completion,
5.4 Certifying completion of the Works, and
7.6 Issuing the final certificate.

These Clauses all cover final decisions concerning acceptance of the Works and payment. It would be against the spirit of naming the Engineer if the Engineer left such decisions to the Resident Engineer. It would also diminish the necessity for the Engineer to have knowledge and control of the day-to-day proceedings of the Site Works.

Delegated authority

2.4. The Engineer or Resident Engineer and/or other suitably experienced person who exercises any delegated power shall upon request of the Contractor specify in writing under which of the foregoing powers any instruction is given. If the Contractor shall be dissatisfied with any such instruction he shall be entitled to refer the matter to the Engineer for his decision.

Delegated authority is only effective when the Contractor has been notified in writing of the delegation by the Engineer. The Employer has no power to appoint the Resident Engineer or any other person, whether 'suitably experienced' or otherwise.

The Contractor is entitled to query with the Engineer any instruction given by those with delegated powers. The intention, once again, is that the Engineer should have intimate knowledge of the day-to-day operations on site so that he is in a position to be able both to check the suitability and cost implications of any instruction given by himself or his staff with delegated authority. This will enable him to certify any 'fair and reasonable costs' incurred additionally by the Contractor as a result of the instruction.

The Contractor is entitled to ask under which clause in the Contract any such instruction is given in order to check that the person giving the instruction has the authority to issue it, and also to get an indication as to whether it is issued under one of the clauses which gives rise to payment or an entitlement to an extension of time.

The powers of the Engineer

Clause 2.3 of the Contract covers all the powers of the Engineer to issue instructions. This is convenient as it removes the necessity to search through the Contract in order to find out the full extent of his powers.

Commentary

The Engineer is given wide powers to give instructions.

2.3. The Engineer shall have power to give instructions for:-

(a) Any variation to the Works including any addition thereto or omission therefrom;
(b) Carrying out any test or investigation;
(c) The suspension of the Works or any part of the Works in accordance with Clause 2.6;
(d) Any change in the intended sequence of the Works;
(e) Measures necessary to overcome or deal with any obstruction or condition falling within Clause 3.8;
(f) The removal and/or re-execution of any work or materials not in accordance with the Contract;
(g) The elucidation or explanation of any matter to enable the Contractor to meet his obligations under the Contract;
(h) The exclusion from the Site of any person employed thereon which power shall not be exercised unreasonably.

These individual powers will now be discussed separately.

Variations to the Works

2.3.(a) Any variation to the Works including any addition thereto or omission therefrom;

The Works are defined in Clause 1.1 as all work necessary for the completion of the Contract including any variations ordered by the Engineer. Since work incorporates temporary works and methods of construction the Engineer's powers are wide enough to enable his to instruct the Contractor to use a specified method or undertake the work in a particular sequence, although such an instruction may have consequences with regard to additional costs and/or extensions of time, and should therefore be given only after careful consideration.

Tests or investigations

2.3.(b) Carrying out any test or investigation;

A power to order tests or investigations is necessary because they cannot be construed as a variation to the Works. The power given to the Engineer is very wide since no definition is given for 'test' or 'investigation'.

Matters covered may include

(i) test for compliance with the Specification of materials or workmanship either before or after inclusion in the Works;
(ii) tests for the suitability of substitute or other materials;
(iii) a final test for acceptance of the Works;
(iv) boreholes, or other site investigations for the benefit of the design, or construction;
(v) reports on any incidents that occur during the execution of the Works; and
(vi) investigations into the condition of existing works or structures.

The sole criteria for payment to the Contractor for tests or investigations is what the Engineer 'considers fair and reasonable' (Clause 6.1). This will include the direct costs incurred by the Contractor in carrying out the test or investigation together with the cost of any delay to the carrying out of the Works that may result. Such delays may include not only those resulting from the carrying out of the tests themselves but also any consequential delays caused by the results of such tests.

The only qualification to payment being made to the Contractor is to be found in Clause 4.4(b) and that is that the Contractor is liable when the test or investigation shows non-compliance by the Contractor with the Contract. This will arise when tests show that materials or workmanship are not in accordance with the Specification or Drawings.

Following the normal *quantum meruit* principle the Contractor should be paid a reasonable percentage addition in respect of profit on the direct costs of the test or investigation, but not on the delay or disruption costs. This is implied by Clause 6.2 (see Chapter 9).

Suspension of the Works

2.3.(c) The suspension of the Works or any part of the Works in accordance with Clause 2.6;

The grounds for suspension are for the proper execution of the work, for the safety of the Works, or by reason of weather conditions. The Engineer is further empowered to issue instructions for the protection of the Works during a suspension.

2.6.(1) The Engineer may order the suspension of the progress of the Works or any part thereof:-
- (a) for the proper execution of the work;
- (b) for the safety of the Works or any part thereof;
- (c) by reason of weather conditions;

and in such event may issue such instructions as may in his opinion be necessary to protect and secure the Works during the period of suspension.

The need for suspension on the grounds of Clause 2.6(1)(a) may arise for matters which are either the responsibility of the Contractor or the Employer. A suspension may be necessary because of matters which are the concern of the Engineer. For instance, when excavation starts on the Site it may reveal information which affects the principles of the design — the bearing capacity of the ground may be lower than that anticipated, the water table may be higher, or services may be exposed which serve adjacent properties.

Alternatively, the Engineer may wish to suspend the Works because the Contractor is executing them incompetently, too slowly for the stability of the ground, or in a manner that is unsafe. In the former case the liability for the costs of the suspension will be for the Employer, and in the latter for the Contractor, since it would be unreasonable for the Employer to bear the costs where the Contractor is at fault (see Clause 6.2).

It should be noted that the power to suspend is limited to factors concerning the 'proper execution of the work', that is, the carrying out of the Works, and does not extend to periods required by the Employer to decide whether or not parts should be omitted, or for the Engineer to alter the design. For obvious reasons (and by inference from Clause 2.6(2)) any Order for

Commentary

Suspension should be in writing, and preferably should state the grounds for the suspension.

2.6(1)(b) for the safety of the Works or any part thereof;

Safety of the Works would seem to imply their structural stability and integrity, but not the manner in which they are executed.

The policy of the Contract is that the Contractor is responsible for the 'adequacy, stability and safety of his site operations'. The Engineer's responsibility is to notify the Contractor, and those involved, if he believes the Works are being executed in an unsafe manner, but to suspend them only if he believes that the Works in their final state will be adversely affected.

It may appear surprising that the Engineer does not have a responsibility under the Contract to suspend the Works if the Contractor's working methods appear to be unsafe. The reason is that the duty to work in a safe manner is imposed by statute and monitoring Contractor's working methods is the responsibility of the Health and Safety Executive (HSE). Nevertheless, everyone concerned with the project has a statutory duty to promote safety and the Engineer might be held criminally liable if unsafe practices could have been known to him but he did nothing.

But that of itself is no reason to order a suspension under the Contract. The better view is that an Engineer who is aware of unsafe working should first alert everyone involved to the danger and, if that has no (or insufficient) effect, he should report the situation to the HSE Inspectorate and then leave it to them to suspend the Works (if that should be necessary) by the issue of a statutory Prohibition Notice. For the Engineer himself to 'jump the gun' and order a contractual suspension where the Works themselves will not be adversely affected might result in his incurring liability in both civil and criminal law, which liability would almost certainly not be covered by his normal professional indemnity insurance.

2.6(1)(c) by reason of weather conditions;

Care should be taken by the Engineer before suspending the progress of the Works due to weather conditions because, unlike other construction contracts, the Contractor is entitled to payment under Clause 6.1 for delays caused by such suspension.

Protection

The power to order protection and security of the Works during suspension is not limited, and the Engineer must specify exactly what he wants because the Contractor has no responsibility for deterioration due to the suspension. Unless the suspension is the direct result of some default of the Contractor, the Employer will have to pay for any protection on a *quantum meruit* basis which will include a reasonable amount in respect of Contractor's profit.

Payment for suspension

The subject of payment generally is discussed in Chapters 9 and 10. However, there are a number of specific points concerning payment for suspensions that should be appreciated.

The Contractor's right to payment is covered by Clauses 6.1 and 4.4(a).

The effect of these two clauses is that the Contractor is entitled to payment for delay and disruption and other additional costs resulting from a suspension ordered by the Engineer provided that he has taken all reasonable steps to avoid or minimize the delay.

This means that the Contractor is entitled to payment unless

(i) the suspension was necessary because of a breach of contract by the Contractor;
(ii) the Contractor could have avoided the necessity for the suspension; or
(iii) once ordered the Contractor fails to take action to remove the cause of the suspension.

Unless one of these applies the Employer is liable for the extra costs incurred by the Contractor as a result of the suspension. The amount to which the Contractor is entitled is that amount which in the opinion of the Engineer is 'fair and reasonable'.

In making this assessment the Engineer must not confuse the two principles liability and amount. He must first decided which party is liable for the costs of the suspension, and in this Contract it is either the Employer or the Contractor. Then he must decided what is a reasonable cost for the extra work done in complying with the suspension order, and any delay or disruption experienced.

Length of suspension.

There must clearly be a limit to the length of time during which the Contractor can be obliged to complete work which is suspended.

> 2.6(2) If permission to resume work is not given by the Engineer within a period of 60 days from the date of the written Order of Suspension then the Contractor may serve a written notice on the Engineer requiring permission to proceed with the Works within 14 days from the receipt of such notice. Subject to the Contractor not being in default under the Contract the Engineer shall grant such permission and if such permission is not granted the Contractor may by a further written notice served on the Engineer elect to treat the suspension where it affects a part of the Works as an omission under Clause 2.3(a) or where the whole of the Works is suspended as an abandonment of the Contract by the Employer.

Such suspension should not exceed 60 days and, if it does, the Contractor is entitled to request permission to start within a further 14 days. If permission is not given the Contractor may then elect to treat the suspension either as an omission of a part of the Works if the suspension only covers a part, or the abandonment of the Works if the suspension is in regard to the whole of the Works.

An Order of Suspension will normally entitle the Contractor to an extension of time provided that the suspension is not his fault and he has taken all reasonable steps to avoid or minimize the delay. The only effect of an extension of time is to relieve the Contractor of payment of Liquidated Damages for lateness under Clause 4.6 and does not necessarily entitle him to additional payment for delay and disruption under Clause 6.1. To qualify for payment not only an extension of time must be awarded under Clause 4.4(a), but also it must be 'fair and reasonable' that payment be made.

The Contractor may serve the written notice requiring permission to proceed any time after the 60 days has elapsed. He does not waive his right by delaying his application.

If the Engineer does not grant such permission within 14 days of receiving the notice it is then up to the Contractor to elect how to proceed. He may

(i) accept a further period of suspension by doing nothing, in which case the Employer will have to pay the Contractor any additional costs incurred as a result of the further delay and disruption.

(ii) where the suspension affects only a part of the work by notice, at any time thereafter treat the work as omitted from the Contract as a variation, in which case the Contractor will be paid the delay and disruption costs up to the time of the notice, and any direct cost resulting from the omission (such as the cost of the providing any shutters not used, or under used, as a result of the omission). The delay costs, if any, will be payable up to the time that the Contractor makes his election, which may be many weeks after the expiration of the 60 day period.

(iii) where the suspension is for the whole of the Works, by notice treat the Contract as having been abandoned by the Employer. The abandonment will take effect from receipt by the Engineer of the Contractor's election notice.

It is doubtful whether the Contractor loses his right to elect to treat the work as an omission under (ii) or as abandoned under (iii) if, subsequent to the expiry of the 14 day notice period, the Engineer then orders the Contractor to proceed. The Engineer would be well advised, therefore, if the Contractor does not make an election, to obtain an undertaking that the Contractor waives his right to do so for a further finite period. In the absence of such an undertaking in the case of (ii) he should issue an order omitting the suspended work to minimize the liability of the Employer.

Change in intended sequence

2.3(d) Any change in the intended sequence of the Works;

The questions which immediately arise are 'what is meant by the *intended* sequence?' and 'whose intention is referred to?' At first sight this seems to compare with the use of the phrase 'specified sequence method or timing' in Clause 51(1) of ICE 5, the adjective 'specified' having been legally determined as meaning 'specified in the Contract' [*Yorkshire Water Authority* v *Sir Alfred McAlpine & Son (Northern) Ltd* (1985)].

If the minor works form draughtsman had meant to follow the ICE 5 analogy he would doubtless have used 'specified'. As he has not, the meaning of 'intended' must be deemed to be wider than 'specified' in ICE 5. The better view is probably that the sequence to be changed may be that necessarily to be implied from the Contract as a whole or that intended by the Contractor when preparing his original tender or programme.

This sub-clause, therefore, gives the Engineer power to control the sequence in which the Works are to be carried out whether that sequence was required by the Contract or merely the Contractor's choice. If it was the latter, and if the Engineer's instruction should lead to delay and/or disruption to the

Works, the Contractor could become entitled to an extension of time under Clause 4.4(a) and/or additional payment under Clause 6.1.

Unforeseen obstruction or condition

2.3(e) measures necessary to overcome or deal with any obstruction or condition falling within Clause 3.8;

Clause 3.8 in the minor works form is the equivalent of Clause 12 in ICE 5.

3.8. If during the execution of the Works the Contractor shall encounter any artificial obstruction or physical condition (other than weather condition or condition due to weather) which obstruction or condition could not in his opinion reasonably have been foreseen by an experienced contractor the Contractor shall as early as practicable give notice thereof to the Engineer. If in the opinion of the Engineer such obstruction or condition could not reasonably have been foreseen by an experienced contractor then the Engineer shall certify and the Employer shall pay a fair and reasonable sum to cover the cost of performing any additional work or using any additional plant or equipment together with a reasonable percentage addition thereof in respect of profit.

For the Engineer to give instructions under this sub-clause it is necessary that the obstruction or condition could not, in his opinion, have been reasonably foreseen by an experienced contractor at the time of formation of the Contract. The 'experienced contractor' is not necessarily the Party to the Contract but a hypothetical contractor skilled in the type of work being undertaken.

Clause 3.8 is discussed in detail in Chapter 6, and 'the unforeseen' is discussed in Chapter 18.

Removal and re-execution of work

2.3(f) The removal and/or re-execution of any work or materials not in accordance with the Contract;

This is the safeguard for poor workmanship or bad materials. Before exercising this power it would be wise for the prudent Engineer to be able to prove that the work rejected was not in accordance with the Contract.

Elucidation and explanation

2.3(g) The elucidation or explanation of any matter to enable the Contractor to meet his obligations under the Contract;

Under ICE 5, Clause 5, any ambiguity or discrepancy between the contract documents or anything which is not clear from the documents is to be 'explained and adjusted' by the Engineer and, if necessary, the Contractor given written instructions. The minor works form Clause 2.3(g) is a simplified version of ICE 5, Clause 5 but is in substance virtually the same in meaning.

Such instructions may be given under this sub-clause, or another, if that sub-clause is more appropriate. The cost effect for the Employer, however, will be similar in either case. The Contractor will be entitled, under Clause 6.1, to any additional cost that he may incur as a result of such instruction. When considering whether additional costs were incurred the Engineer will have to decide what the Contractor should have allowed for in his tender on the basis of the unclear documents. If in doubt he should remember that the responsibility for the accuracy of the documents rests with the Employer.

Exclusion of persons

2.3(h) The exclusion from the Site of any person employed thereon which power shall not be exercised unreasonably.

The Contract gives no guidance as to the grounds upon which persons can be excluded. This is sensible in a contract of short duration, but Engineers should exercise the power with care since any costs necessarily incurred by the Contractor as a result of an exclusion instruction would have to be paid for by the Employer if such powers were exercised unreasonably.

Guidance on the intention behind this sub-clause can be gained by reference to Clause 16 of ICE 5. That gives the Engineer power to require the removal of a person who

(i) misconducts himself,
(ii) is incompetent or negligent in the performance of his duties,
(iii) fails to conform with any particular provisions with regard to safety which may be set out in the Specification, or
(iv) persists in any conduct which is prejudicial to safety or health.

Where the Engineer has powers to give instructions for matters which are not either the fault or the responsibility of the Contractor then the Contractor is entitled to additional payment for the additional costs involved under Clause 6.1, and for the costs of delay and disruption, including but not limited to, the matters referred to in Clause 4.4.[1]

Daywork

2.5. The Engineer may order in writing that any work shall be executed on a daywork basis. Subject to the production of proper record the Contractor shall then be entitled to be paid in accordance with a Daywork Schedule included in the Contract or otherwise in accordance with the 'Schedules of Dayworks carried out incidental to Contract Work' issued by the Federation of Civil Engineering Contractors current at the date the work is carried out.

There are three alternative ways of specifying a daywork schedule, either by a contract schedule drawn up in accordance with the method of measurement used for the preparation of the Bills of Quantities, or by specifying that the conditions contained in the 'Schedule of Dayworks carried out incidental to Contract Work' issued by The Federation of Civil Engineering Contractors current at the date of the tender shall be used, or by providing for the Contractor to quote percentage adjustments to the rates contained in the Daywork Schedule. The basis of this schedule and its use is discussed in Chapter 16, and examples of each are given in Appendix E.

If a schedule for the Contract is to be drawn up it is essential that it should clearly define what is to be included in the rates. Usually, the method of measurement will define the parameters. Such items as productivity bonus,

1. See Clause 2.4 on p. 33.

overtime rates, wastage of materials, cost of supervision, cost of general site transport and general site overheads all lead to problems if the manner in which they are to be covered is not clearly specified somewhere in the Contract.

It is also necessary to state the extent to which the rates are to apply. The FCEC schedule is specifically designed for works 'incidental to the contract', and will not usually cover the whole of the site overheads if the only work being carried out is valued at normal daywork rates.

If the FCEC daywork schedule is to be used two approaches are possible. The schedule is divided into three sections for labour, materials and plant. Each section has a different percentage addition applied to the rates, or prices in it to cover the Contractor's overheads and profits. The FCEC publishes suggested percentage additions. These additions can either be specified to be applicable in the contract documents or the Contractor can be asked to quote separately the percentage additions that he requires.

Where the Daywork Schedule is required only to value incidentals, then it is probably appropriate to specify that the FCEC percentages shall be applicable, in which case the Bill of Quantities will only contain a provisional sum for the daywork items. However, if the entire Contract is to be valued on the basis of daywork rate then, in order to obtain some element of competition into the tendering process, the Engineer should assess the likely expenditure on the proposed work for labour, materials and plant as lump sums and invite the Contractor to add his own percentage additions to these lump sums. The total of the lump and the percentage sum additions will then be the total of the Contractor's offer.

The three different methods of billing dayworks are given in Chapter 16.

Parties bound by Engineer's instructions

2.7. Each party shall be bound by and give effect to every instruction or decision of the Engineer unless and until either:-

(a) it is altered or amended by an agreed settlement following a reference under Clause 11.3 and neither party gives notice of dissatisfaction therewith, or
(b) it is altered or amended by a decision of an arbitrator under Clause 11.4 or 11.5.

This clause gives efficacy to the Contract and enables the work to continue if one of the parties does not agree with either the other party or the Engineer. The Engineer issues an instruction to resolve the difficulty, which both parties are obliged to carry out. The parties can always refer the matter to conciliation or arbitration later if they disagree with the Engineer's instruction (see Chapter 14.)

Comparison with ICE 5

The Institution of Civil Engineers suggests that the Engineer under ICE 5 should be a named person but it is not a requirement, whereas under the minor works form it is.

Unlike ICE 5 the minor works form collects all the powers of the Engineer together under one clause — Clause 2.3. This is a convenient arrangement.

Commentary

Under both contracts the Engineer has a central role in producing the design, supervising the construction, and certifying payment and completion.

The power to delegate is greater under the minor works form, although the Notes for guidance suggest certain limitations that are not mandatory. The Engineer can delegate all his powers to the Resident Engineer, whereas under ICE 5 the power is restricted in relation to final decisions: final certificate, completion certificates, and extension of time awards.

The provisions for testing and making investigations are collected together in the minor works form whereas they are distributed throughout ICE 5 in Clauses 13, 18, 36, 37, 38 and 50.

Powers that the Engineer has by express provision under ICE 5 but not under the minor works form are

- (i) to approve a programme (Clause 14(1)),
- (ii) to consent to temporary work (Clause 14(3)),
- (iii) to approve Contractor's representative (Clause 15(2)),
- (iv) to instruct safety facilities, except as variations (Clause 19(1)),
- (v) to instruct as to how fossils or antiquities are to be dealt with (Clause 32),
- (vi) to give permission for night or Sunday work (Clause 45),
- (vii) to expedite progress (Clause 46),
- (viii) to nominate sub-contractors (Clause 58),
- (ix) to certify Contractor's default (Clause 63), and
- (x) to settle disputes, although Clause 2.7 gives interim power, subject to later conciliation or arbitration (Clause 66).

Table 1 details the powers of the Engineer under ICE 5, and shows how they are embodied into the minor works form.

Clearly, the presence or absence of these powers will influence the proper choice of contract. When deciding which form of contract to use the Engineer should note the following points.

- (i) Since the minor works form contains no power or requirement that the Contractor's programme be approved, any modification which the Engineer may require can only be enforced by an instruction (see Clause 2.3(d), page 38). Such an instruction may well give rise to a claim for additional payment. Provided, however, that design details are available at tender stage and the Site is available on time at the starting date, this should in practice lead to few problems.
- (ii) Similarly, the Engineer has no express power to request details of the Contractor's temporary works. Should this be likely to cause difficulty, ICE 5 should be used.
- (iii) The absence of any power to issue instructions on safety provisions means that any necessary requirements must be detailed in the Specification.

The Contractor is of course bound by the law in regard to safety requirements, but should the Engineer require safety facilities additional to the minimum legal standard, this can subsequently only be obtained by issuing a variation under Clause 2.3(a) (see also the discussion under Clause 2.6(a), page 35).

(iv) There is no power to suspend the Works if fossils or antiquities are discovered. The only course of action open to the Engineer is to try to procure an agreement between Employer and Contractor on how the fossils are to be dealt with, or to make suitable provision in the Specification.

(v) Night or Sunday working is not proscribed by the Contract. If the Contractor is to be forbidden to carry out work at night or on Sundays then such a clause should be written into the Specification. Alternatively, such a provision could be included in the Contract as a special condition. If such a clause is included, it should start with the words 'unless the Engineer shall instruct otherwise, no work may be executed at night or on Sundays . . .'. The omission then will have no effect.

(vi) The lack of power to order the Contractor to expedite progress is discussed in Chapter 7.

(vii) The lack of provision for nominated sub-contractors is discussed in Chapter 11.

(viii) The lack of a default clause is discussed in Chapter 7.

(ix) The minor works form has introduced optional conciliation (see Chapter 14) in place of the reference to the Engineer for a formal decision under Clause 66(1) if ICE 5 as the first stage in resolving disputes. However, as in ICE 5, both the Contractor and the Employer are bound to give effect to any instructions of the Engineer pending final resolution of the dispute. Under ICE 5 the Contractor was bound to comply by Clause 13(1) and it is necessarily implied that the Employer will do so. Clause 2.7 of the minor works form now provides for this in express words.

Where the risks inherent in the Contract are judged to be small it is likely that any disputes that do arise will result from differing opinions between the Engineer and the Contractor on the interpretation of the contract documents or instructions given by the Engineer. In most cases such disputes should be resolved by seeking the view of an independent person in whom both the parties have confidence, and that is the aim of conciliation.

Comparison with ICE 6

ICE 6 now contains express provisions analogous to those in the minor works form with regard to the 'named individual' as Engineer, the collection of the Engineer's powers into one clause and the requirement that both parties comply with the Engineer's instructions and decisions.

The Engineer is given the express power to agree that any land used of the purpose of the Works shall be designated as part of the Site. He is also now permitted to give any instruction orally that 'it is considered necessary to give . . . orally' and not just orders for variations as in ICE 5.

For the remainder, the comparison with ICE 6 is broadly the same as that with ICE 5, except that ICE 6, Clause 66 now contains both the condition precedent to arbitration that a dispute be referred back to the Engineer for

Table 1. Comparison of Engineer's powers under ICE Conditions of Contract and minor works form

	Power under ICE 5 and 6	Clause	Clause	Comment on minor works form
1	Appointing an Engineer's Representative (ER)	1(1)(d)	2.2	The Engineer may appoint any suitably experienced person to assist him, and there is no restriction on his power to delegate. Additional obligation to identify clause under which power arises
2	Appoint assistant to the ER	2(2)	2.2	
3	Delegate power to assistants	2(3)	2.2	
4	Consent to sub-letting	4	8.2	Consent not to be unreasonably withheld
5	Correct ambiguities and discrepancies	5	2.3(g)	
6	Issue revised and additional drawings	7(1)	2.3(a)	Variations only
7	Issue instructions concerning adverse physical conditions and artificial obstructions	12(2)	3.8(a)	
8	Issue instructions and directions	13(1)	3.8	No specific general power
9	Approve a programme	14(1)	4.3	No power to approve only request modifications
10	Request a revised programme	14(2)	4.3	
11	Request details of Temporary Works	14(3)		No power to request details
12	Give consent to methods of construction	14(4)		No requirement
13	Approve Contractor's representative	15(2)		No power or requirement
14	Order removal of persons from Site	16	2.3(h)	Not to be exercised unreasonably
15	Instruct boreholes or exploratory excavations	18	2.3(b)	All tests covered by same power (see 19 and 50)
16	Instruct provision of safety facilities	19(1)		No power. Such facilities must be detailed in specification, or ordered as variations
17	Instruct facilities for other contractors	31(1)	3.9	Payment under Clause 6.1
18	Instruct on fossils or antiquities	32		No power
19	Instruct tests on materials and workmanship	36(1)	2.3(b)	See item 15
20	Order removal of materials	39(1)(a)	2.3(4)	General instruction to remove and re-execute work
21	Order substitute materials	39(1)(b)	2.3(4)	
22	Order removal and re-erection of constructed work	39(1)(c)	2.3(4)	

No.	Power	Clause	Clause	Comment
23	Suspend the Works	40	2.6	No power to suspend work before it starts
24	Order commencement	41	4.1	Restricted to within 28 days after acceptance
25	Award extensions of time	44	4.4	Similar.
26	Give permission for night or Sunday work	45		No power. The Contract rules
27	Order steps to expedite work	46		No power
28	Issue Certificate of Completion	48(1)	5.4	
29	Order searches and tests for defects	50	2.3(b)	See item 15
30	Order variations to the Works	51	2.3(a) 2.3(d)	
32	Value variations	52(1)	6.2	Simpler rules. Only criterion is what is 'fair and reasonable'. Regard only to be given to tender rates
33	Fix rates	52(2)	5.1	
34	Order work to be on a daywork basis	52(3)	2.5	
35	Value claims	52(4)	6.1	No notices required except of 3.8
36	Correct errors in Bill of Quantities	55(2)	2.3(g)	General power to explain
37	Measure and value the Works	56(1)	7.3	Several different methods of payment are permitted
38	Order work to be carried out under Provisional Sums	58(1	2.3(a)	Variations only
39	Order work to be executed by Nominated Sub-contractors	58(4)	—	No nominated sub-contractors
40	Certify interim payments	60(2)	7.3	Similar
41	Issue final certificate	60(3)	7.6	Tighter timetable for issue
42	Omit value of work previously certified	60(7)	7.3	Implied power
43	Issue Maintenance Certificate	61(1)	5.4	Called 'Defects Correction Certificate'
44	Certify Contractor's default	63		No default clause
45	Settle disputes	66	11	Conciliation substituted for Engineer's decision, but Clause 2.7 give interim power to decide

a formal decision and a provision for optional conciliation, using virtually the same words as in the minor works form.

Legal comments

Legal position of the Engineer

1. *Relationship between Engineer and Employer*

Consulting Engineers are normally engaged on the ACE Terms of Engagement, whereas Engineers who are the employees of the Employer, such as local government officials, will be employed under the terms of their contracts of employment. In both cases, their duty when carrying out their functions under a construction contract is to the Employer, but that duty is imposed by different means for the two different types of person.

The Consulting Engineer is required both by the ACE Conditions and the law to exercise all reasonable skill, care and diligence in the discharge of his duties and, also, when exercising a discretion as between the Employer and the Contractor, he is required to act fairly.

Statute law also imposes duties upon him, in the absence of a specific contractual term. The Supply of Goods and Services Act 1982, implies to all contracts a term that the services will be carried out with reasonable skill and care and within a reasonable time. Again, where there are specific contractual terms governing such matters, it has been held that these will involve at least due care and diligence [*Lister* v *Romford Ice and Cold Storage Company* (1957) AC 555]. Attempts have also been made to impose a strict liability for fitness for purpose by express warranty or implication [*Greaves* v *Baynham Meikle* (1972) 1 WLR 1095].

What is remarkable, however, is that the normal terms of engagement do not attempt, other than in the general sense above, to define the limits of responsibility and liability of the Engineer. Liability for design is also generally limited to the use of reasonable professional skill and care, although fitness for purpose has been implied in a number of cases. However, these latter cases have been limited to those in which the design or the selection has been accompanied by the provision of materials and/or the construction of the Works [*Young and Martin* v *McManus Childs* (1969) 1 AC 454, *IBA* v *EMI*, 14 BLR 1, *Viking Grain* v *White*(1985) 33 BLR 103].

Liabilities in tort

The law of tort is changing rapidly as a result of a number of recent cases [*D&F Estates* v *Church Commissioners* (1988) 3 WLR 368, *Greater Nottingham Co-operative Society* v *Cementation Foundations and Engineering* (1988) 41 BLR 43, *Glenlion Construction Ltd* v *The Guinness Trust* (1987) 30 BLR 50, *Murphy* v *Brentwood District Council* (HL 1990)]. However, the following points should be noted:

(i) The courts will generally not impose tortious duties when the parties have chosen not to impose such duties in their contract.

(ii) Liability in tort will arise generally only where there is physical damage and not just financial (pure economic) loss.

(iii) Gratuitous advice which leads to financial harm will normally permit a direct claim for negligent mis-statement if there is a special relationship between the parties [*Hedley Byrne* v *Heller and Partners* (1964), AC 465]. When certifying, the Engineer has a duty of care to the Employer [*Sutcliffe* v *Thackrah* (1974) 1 All ER 859].

2. *Engineer as agent of the Employer*

Following the decision in *Pacific Associates* v *Baxter and Others* [44 BLR 33], it was decided that in everything the Engineer does under the FIDIC Contract (which is based upon ICE 5), he acts as agent for the Employer. An agent under the law is a person who has authority to enter into contracts on behalf of another person, his principal. Once the contract is made, then the agent drops out and normally retains no liability for its performance.

An agency may arise

(i) by express agreement with the principal (this is known as express authority), or
(ii) by implication in order to enable the agent to perform his task, or
(iii) if the principal treats the agent as having that authority (this is known as apparent or ostensible authority).

When an agent acts within his authority, then the principal is liable. If he acts outside that authority, however, he will be liable to the third party for damages on the basis of an implied warranty of authority. Under most construction contracts, for much of the time the Engineer (or Architect) will be acting as agent of the Employer. The Contract itself effectively sets out the authority of the Engineer, but his actual agency agreement contains the terms under which he is engaged by the Employer.

The effect of agency arrangements as between Employer, Engineer and Contractor will generally be implied as follows:

(i) between Employer and Engineer, that the Engineer has authority to act as the conditions say he shall, and
(ii) between Employer and Contractor, that the Contractor is entitled to act on the basis that the Engineer has apparent authority to exercise all of the powers given to him by the conditions.

This means that if the Employer imposes limitations on the powers of the Engineer, unknown to the Contractor, and the Engineer exceeds these powers, then he will be liable to his employer in damages, but the Employer will be unable to avoid liability under the Contract to the Contractor for those actions of the Engineer which are in excess of his actual authority.

Duties imposed upon the Engineer by law

When the Engineer acts as agent for the Employer, the law implies that he will do so honestly [*Neodox* v *The Borough of Swinton and Pendlebury* (1958) 5 BLR 34].

When acting as certifier, he is required to act fairly as between the parties. It is implicit in these arrangements that the Employer will not interfere with the Engineer issuing a certificate, nor should he interfere with other essential functions, such as the issuing of variation orders or the granting of extensions of time [*Hickman & Co.* v *Roberts* (1913) AC 299, *Perini Corporation* v *Commonwealth of Australia* (1969) 12 BLR 82, *Holland Hannen and Cubitts (Northern) Limited* v *Welsh Health Technical Services Organisation* (1985) 35 BLR 1].

3. Relationship between Engineer and Contractor

The Engineer has no contractual liability to the Contractor, since there is no contract between them, and the doctrine of privity prevents the Contractor from suing the Engineer for damages resulting from wrongful instructions under the construction contract. Any liability that the Engineer has will therefore arise in tort. Following the decision in *Pacific Associates* v *Baxter*, it is unlikely that the Contractor would succeed in a claim against the Engineer for any wrongful action made in the exercise of his duties.

In the course of his Judgment in that case, Lord Justice Gibson said

> The contractual duty of the Engineer, owed to the Employer, to act fairly and impartially is a duty in the performance of which the Employer has a real interest. If the Engineer should act unfairly to the detriment of the Contractor, claims will be made by the Contractor to get the wrong decisions put right. If arbitration proceedings are necessary, the Employer will be exposed to the risk of costs in addition to being ordered to pay the sums which the Engineer should have allowed. If the decisions and advice of the Engineer, which caused the arbitration proceedings to be taken, were shown by the Employer to have been made and given by the Engineer in breach of the Engineer's contractual duty to the Employer, the Employer would recover his losses from the Engineer.

The position seems to be, therefore, that any wrongful act of the Engineer will put the Employer into breach and the Contractor will have a remedy under the Contract. The Employer may then have an action against the Engineer for breach of duty.

4. Collateral warranties

A collateral contract is a contract which exists side by side, or is subordinate to another contract. It may impose additional liabilities on the parties to the original contract, or it may seek to pass the benefits and obligations of the original contract to third parties. The latter circumvents the doctrine of privity. Such a contract in relation to a contract for the supply of services which seeks to guarantee the performance of those services, either to one of the original parties, or to third parties, is called a collateral warranty.[1]

A practice has recently arisen by which Employers seek to extract from their Engineers written collateral warranties aimed at imposing strictly liability (i.e. regardless of fault) for the success of a project, and guarantees of fitness for purpose in respect of design work.

Such warranties are also often drafted so as to benefit third parties, such

---

1. Further information can be obtained from *Collateral warranties — a practical guide for the construction industry* by David L. Cormes and Richard Winyard (1990). Oxford: Blackwell Scientific.

as future and as yet unknown tenants. This is particularly pernicious, since unless the identity and occupation of a future tenant is known, there will often be no basis on which to judge 'fitness for purpose'.

These matters are outside the scope of this book, but all such warranties should be resisted if at all possible. If commercial circumstances are such that a warranty of this kind cannot be avoided, it should be covered so far as may be possible by appropriate insurance, with the cost of that insurance added to the fees and disbursements of the Engineer. Insurers should always be consulted before signing any collateral warranty, as it is possible to invalidate existing insurance policies by accepting additional liabilities without the acceptance of them by the insurer.

Chapter 6

GENERAL OBLIGATIONS

Introduction

The Contractor's general obligation is to supply everything and do everything necessary to complete the Works. The Contract tends to restrict this general obligation in specific ways.

The Engineer is responsible for providing all details and information necessary to enable the Contractor to fulfil his obligations. The Contractor is not normally responsible for the design of the permanent works.

Obligation to perform and complete

3.1. The Contractor shall perform and complete the Works and shall (subject to any provision in the Contract) provide all supervision labour materials plant transport and temporary works which may be necessary therefor.

By accepting an undertaking to complete the Works the Contractor implicitly warrants that the Works are capable of being completed, even though the design and specification of those Works is otherwise the Employer's responsibility.

The Contractor is responsible for the design and supply of temporary works. Therefore, during the tender period he must satisfy himself that he has the expertise to complete the Works, that the Works as designed can be completed and that his proposed methods are feasible. This is the total extent of the Contractor's liability for design.

The Contractor must provide supervision but there is no requirement for that supervision to be constantly upon the Site. On small contracts tradesmen and others may be expected to work either singly or in small gangs and supervise themselves.

The words in brackets 'subject to any provision in the Contract' cover materials and plant to be supplied by the Employer, and presumably the supply of goods and services by sub-contractors named in the Contract.

Care of the Works

3.2. (1) The Contractor shall take full responsibility for the care of the Works from commencement until 14 days after the Engineer issues a Certificate of Practical Completion for the whole of the works pursuant to Clause 4.5

(2) If the Engineer issues a Certificate of Completion in respect of any part of the Works before completion of the whole of the Works the Contractor shall cease to be responsible for care of that part of the Works 14 days thereafter and the responsibility for its care shall then pass to the Employer.

(3) The Contractor shall take full responsibility for the care of any outstanding work

which he has undertaken to finish during the Defects Correction Period until such outstanding work is complete.

Care of the Works means protecting them from damage howsoever arising. This will include damage from fire, storm, tempest, earthquake, damage due to the Contractor's operations or his sub-contractors', and damage due to trespassers. However, most of these hazards are insurable and are discussed in Chapter 11.

Commencement is the date from which the Contractor is given access to the Site and he is responsible for the Works from then until 14 days after the Engineer issues a Certificate of Practical Completion. The 14 days' overrun is to enable the Employer to be able to take out his own policy of insurance if he so wishes.

The Works can be taken over either as a whole or in parts according to the manner in which the Appendix to the Conditions of Contract has been completed. Completion is discussed in Chapter 7 but it should be noted that the Contractor's responsibility for the care of the Works under Sub-clause (2) also ceases in respect of a part of the Works 14 days after it is taken over.

Finally, even after the issue of the Completion Certificate, the Contractor will often be obliged to do further work, either in completing works still outstanding when the certificate is issued or to put right defects which appear during the Defects Correction Period. It is not clear whether the Engineer can issue a variation order after the Completion Certificate has been issued and before the end of the Defects Correction Period, but provided that the Contractor is prepared to undertake the work this should present no problem.

For obvious reasons, the Contractor will continue to be responsible for all such outstanding, remedial or varied work until such time as it has been completed, but this residual responsibility will not extend to parts of the Works which are not affected by such work.

Repair and making good

3.3. (1) In case any damage loss or injury from any cause whatsoever (save and except the Excepted Risks) shall happen to the Works or any part thereof while the Contractor is responsible for their care the Contractor shall at his own cost repair and make good the same so that at completion the Works shall be in good order and condition and conform in every respect with the requirements of the Contract and the Engineer's instructions.

(2) To the extent that any damage loss or injury arises from any of the Excepted Risks the Contractor shall if required by the Engineer repair and make good the same at the expense of the Employer.

(3) The Contractor shall also be liable for any damage to the Works occasioned by him in the course of any operations carried out by him for the purposes of completing outstanding work or complying with his obligations under Clauses 4.7 and 5.2.

As a consequence of the Contractor's 'full responsibility' for the care of the Works under Clause 3.2 he is also required to make good any damage at his own expense. Normally this liability will be covered by the insurance of the Works. However if the damage is caused by the Excepted Risks then such damage is made good at the expense of the Employer. The most important Excepted Risks are

(i) the use or occupation of the Works by the Employer,
(ii) damage by other Contractors (not employed by the Contractor),
(iii) an error in the design (not provided by the Contractor).

These items are not covered by the insurance policy for the care of the Works but should be covered by other insurance policies taken out by or on behalf of the Employer.

Sub-clause (1) enables the Engineer to give instructions as to how damaged work shall be made good. If such instructions require the Works to be made good to a standard better than that provided for by the Contract then such instructions would constitute a variation for which the Contractor would be entitled to additional payment.

The Contractor would only be entitled to an extension of time under Clause 4.4 if the cause of the damage was 'outside the control of the Contractor'.

The Contractor may be required to carry out certain works after the issue of the Certificate of Practical Completion. These works could include making good defects and carrying out any work outstanding at the time of the issue of the Certificate. Whilst carrying out these items the Contractor remains responsible for them. Again, any damage caused will be covered by the policy of insurance for the Works. These matters are covered by Clause 4.7 (Rectification of defects) and Clause 5.2 (Cost of remedying defects).

Contractor's authorized representative

3.4. The Contractor shall notify the Engineer of the person duly authorized to receive instructions on behalf of the Contractor.

On smaller sites, the Contractor may not have a full-time representative on the Site, and the person authorized to receive instructions will be based at his office. It may on occasion be necessary, therefore, for the Engineer to give instructions to somebody other than the nominated person (for instance, a sub-contractor on site) and this, although not covered by the Conditions of Contract, is envisaged in the Notes for guidance.

NG11. In respect of Clause 3.4, it has to be recognized that in a minor works contract, the Contractor might have no full-time supervisor on site, and the Contractor may ask for instructions to be delivered or sent elsewhere for the attention of his representative. In these circumstances, the Contractor has to accept the fact that urgent instructions might, in the interests of safety, or for some other reason, have to be given directly to the Contractor's operatives on site.

The Contractor's representative on site may well be a foreman, operative, or even an employee of a sub-contractor. A flexible approach to instructions must be maintained, and the Engineer should, if he gives instructions other than to the Contractor's representative, immediately confirm such instructions in writing to the representative, if relations are not to break down. On sites where it is known that there will be no full-time representative, clear arrangements for the receipt of instructions should be made between the Engineer and the Contractor's representative before work starts.

Setting out and safety

3.5. The Contractor shall take full responsibility for the setting-out of the Works and for the adequacy stability and safety of the site operations and methods of construction.

The Contractor is fully responsible for all site operations. There is no contractual obligation placed upon the Engineer to check the setting-out or approve the methods of construction. He may choose to do so in the interests of the Employer but, if he does, it will not relieve the Contractor of any of his responsibilities under the Contract.

Adequacy and stability refer to both the permanent and the temporary works. The Contractor must ensure that his temporary works are adequate to produce the quality required by the Specification, and this means both in finish and tolerance.

The consideration of safety before any construction operation is commenced is essential. The construction industry has an unenviable record with regard to safety. The Contractor, under the Contract, is made fully responsible for the safety of all site operations. Thus he bears the principal duty to comply with the Health and Safety At Work Act, such as appointing a site safety supervisor and maintaining records. However, this will not necessarily relieve either the Employer or Engineer of any duty with regard to safety which may be laid upon them by statute.

The Engineer is responsible for the safety of his own staff, and has a duty (but not under the Contract) towards the Contractor, and any others whom he considers may be affected, to inform them of anything which he perceives to be a danger.

Responsibility for design

3.6. Subject to Clause 3.5 the Engineer shall be responsible for the provision of any necessary instructions drawings or other information.

3.7. The Contractor shall not be responsible for the design of the Works except where expressly stated in the Contract. The Contractor shall be responsible for the design of any temporary works other than temporary works designed by the Engineer.

These two clauses need to be read together.

Since the design is meant to be complete at the time of tender, the Engineer should be in a position to supply full constructional details at the start of the Contract. He will remain responsible for supplying the necessary drawings to enable the Contractor to complete his obligations under the Contract. When the Contractor requires further drawings, or other information, it will be up to him to request it as soon as the need becomes apparent. To obtain additional payment under Clause 6.1 for failure of the Engineer to supply the necessary details in time, the Contractor will have to show that he requested the information as soon as the need became apparent. It is, however, still the Engineer's duty to supply all the necessary information and the Contractor is not responsible for checking that he has it.

The Contractor is not normally responsible for the design of the permanent works unless the Contract provides otherwise.

NG6. If the Contractor is required to be responsible for design work of a specialist nature

> which would normally be undertaken by a specialist sub-contractor or supplier (such as structural steelwork, mechanical equipment or an electrical or plumbing installation) full details must be given either in the Specification or in the Appendix to the Conditions of Contract or on the Drawings indicating precisely the Contractor's responsibility in respect of such work.

Such an express statement may be included when there are specialist Works for which the Contractor is tendering. For instance, if the Works are to provide piles for a foundation, many specialist piling contractors undertake their own design, and in such cases, that part of the design may be an express requirement of the Contract.

An express statement must be clear beyond all doubt. To be certain that the design is included in the Contract, the Engineer should, in the Specification, use the words 'the Contractor shall take full responsibility for the design of . . .'.

The Contractor is responsible for the design of temporary works, unless they are designed by the Engineer. This responsibility will include obtaining from the Engineer all necessary information required to enable the temporary works to be designed. A delay in asking for such information will not entitle the Contractor to either more time or additional money. However, if the Engineer is slow in providing such information, then the Contractor may well be entitled to both.

Adverse physical conditions and artificial obstructions

> 3.8. If during the execution of the Works the Contractor shall encounter any artificial obstruction or physical condition (other than a weather condition or a condition due to weather) which obstruction or condition could not in his opinion reasonably have been foreseen by an experienced contractor, the Contractor shall as early as practicable give a written notice thereof to the Engineer. If in the opinion of the Engineer such obstruction or condition could not reasonably have been foreseen by an experienced contractor, then the Engineer shall certify and the Employer shall pay a fair and reasonable sum to cover the cost of performing any additional work or using any additional plant or equipment together with a reasonable percentage addition in respect of profit as a result of:
>
> (a) Complying with any instructions which the Engineer may issue
>
> and/or
>
> (b) taking proper and reasonable measures to overcome or deal with the obstruction or condition in the absence of instructions from the Engineer
>
> together with such sum as shall be agreed as the additional costs to the Contractor for the delay or disruption arising therefrom. Failing agreement to such sums the Engineer shall determine the fair and reasonable sum to be paid.

The policy behind the inclusion of Clause 3.8 in the Contract is that the tenderer shall not be required to include an allowance in his tender price for conditions or obstructions which may occur upon the Site, but which he could not reasonably foresee. If contractors had to include for such matters, the general level of prices for civil engineering works would need to be increased substantially to cover for this risk. It follows therefore that the risk of such unforeseen conditions or obstructions being encountered falls upon the Employer.

If, during the execution of the Works . . . .

The execution of the Works, which includes both the permanent and the temporary works, will be on the Site, so what follows refers to matters experienced upon the Site.

. . . any artificial obstruction or physical conditions . . .

What constitutes artificial obstructions or physical conditions is discussed in Chapter 18 but both are physical impediments to the progress of the Contractor's operations.

. . . (other than a weather condition or condition due to weather) . . .

As with artificial obstructions or physical conditions, the weather conditions must be those experienced upon the Site. Such matters as extreme cold, extreme heat, unusually dry or wet conditions, the results of storms and high winds, will all be excluded from this condition, but the results of storms fifty miles away which cause flooding may well come within its scope. In case of doubt, each situation must be considered on its merits.

. . . could not in his opinion reasonably have been foreseen by an experienced contractor . . .

An 'experienced contractor' is not necessarily the one undertaking the Works. It is a hypothetical contractor who is experienced in the type of construction being undertaken and therefore expected to be able to anticipate the effects of difficulties normally encountered in that type of work. Since this is a common standard for all tendering contractors, all tenders are therefore prepared on the same basis.

. . . could not reasonably have been foreseen . . .

The question must be related to the particular Site. For instance geological faults are widespread in the United Kingdom, and there is always the possibility of finding one on any site. If one is found the question to be asked is 'why should I expect a fault on this site?'

If there is no evidence of the presence of a fault then the Contractor could not reasonably have foreseen it when one is found. When assessing foreseeability, therefore, the Engineer must examine what is found and then ask himself the question 'could I have expected to find this particular condition?' If he can provide the evidence indicating its presence from topographical features of the Site, or from records that the Contractor could reasonably have been expected to discover during the tender period then the Contractor could have foreseen it, otherwise he could not.

. . . the Contractor shall . . . give written notice . . .

The Contractor is required to give notice as early as practicable upon encountering any such obstruction or condition. It is envisaged that the Engineer will not have full-time supervision upon the Site and will rely upon the Contractor to tell him when the unforeseen conditions or obstructions are found. The notice will alert the Engineer and enable him to inspect the Site prior to the obstructions being overcome, so that, if necessary, he can issue instructions or, in suitable cases, even vary the Works.

Another reason that it is important for the Contractor to give notice as quickly as possible is that if the Engineer is unable to check upon the effect of the obstructions or conditions, he will be unable to certify money in respect of additional costs or grant an extension of time for any delays encountered.

Action on receipt of notice

The Contractor is entitled to be paid for dealing with obstructions or physical conditions whether he is instructed to do so or not. The Engineer therefore should inspect the Works quickly upon receipt of the notice from the Contractor so that he may change or stop the measures that the Contractor is taking if he disagrees with them, and give instructions as to how he wishes them to be overcome.

For his part, the Contractor should not wait for the Engineer to react to his notice, since he can and should consider and put in hand such proper and reasonable measures to deal with the condition or obstruction as may be available to him. It is only if the Engineer intervenes with an instruction or variation that the Contractor will be relieved of this need to exercise self-help. It is for this reason that Clause 3.8 contains the provision for the Employer to pay the Contractor for

> (a) Complying with any instructions which the Engineer may issue
>
> and/or
>
> (b) taking proper and reasonable measures to overcome or deal with the obstruction or condition in the absence of instructions from the Engineer.

Indeed, the draughtsman of the minor works form could perhaps be criticized for not putting these two alternatives the other way round.

It should also be remembered that, whenever a claim arises, the Claimant is by law under an implied duty to 'mitigate damage', that is, to keep the sum involved to a minimum. The Contractor must therefore strive to keep both the additional cost and the delay which arises from the condition or obstruction to a minimum. It must follow that, if it is readily apparent how the problem should be dealt with, the Contractor should get on and do it. If, however, the solution is not reasonably obvious he should not hesitate to press the Engineer for instructions.

Payment

The Contractor's entitlement to payment (if any) comes under two headings. Firstly, if he does additional work or uses additional equipment he is entitled to the reasonable cost for that work or equipment together with a reasonable percentage of profit.

> . . . If in the opinion of the Engineer such obstruction or condition could not reasonably have been foreseen by an experienced contractor then the Engineer shall certify and the Employer shall pay a fair and reasonable sum to cover the cost of performing any additional work or using any additional plant or equipment together with a reasonable percentage addition in respect of profit . . .

Additional work is work that would not have been done but for the encountering of the obstruction or condition and additional plant means plant that would not have been used for the operation if the obstruction or condition had not been found. The slowing down of the existing resources that were there when the condition was found is 'delay or disruption' and comes under the second heading.

The second heading is for delay and disruption for which the Contractor is entitled to any additional cost incurred. These two headings follow the normal rule in the Contract that additional work attracts profit, whereas delay and disruption only entitles the Contractor to additional costs.

> . . . together with such sum as shall be agreed as the additional costs to the Contractor for the delay or disruption arising therefrom. Failing agreement to such sums the Engineer shall determine the fair and reasonable sum to be paid.

The Contractor and the Engineer should, in consultation, agree the additional costs involved and for this purpose the Contractor will need to produce records of his output before, during and after meeting the obstructions together with the resources involved at all times. The Contractor, therefore, should as soon as he finds such conditions or obstructions do two things. Firstly he should inform the Engineer of what has been found and its likely effects, and secondly he should make sure that he takes records of the costs involved.

If the Engineer and the Contractor cannot agree the costs, then the Engineer is to determine what is fair and reasonable. If the Contractor disagrees with such an assessment then it is open to him to challenge the figure in arbitration or, preferably, by the parties adopting the conciliation option. In addition to payment the Contractor may also be entitled to an extension of time. (See Chapter 7.)

Facilities for other contractors

> 3.9. The Contractor shall in accordance with the requirements of the Engineer afford reasonable facilities for any other Contractor employed by the Employer and for any other properly authorized authority employed on the Site.

While the Contractor may well have his own sub-contractors on Site, there will also be occasions when entities other than such sub-contractors are also on Site. These may include public utilities (whether having work to do in connection with the Employer's project or otherwise) or other contractors directly employed by the Employer. In such situations the Contractor will not only be sharing the Site with them but may be required to afford them reasonable facilities, such as giving the other contractors access to parts of the Site with which they are concerned, making suitable areas of the Site available for offices and storage of material, supplying power and water and the like.

Unless the Specification or the Contract require facilities of a special nature to be provided by the Contractor, the Contractor's liability will be limited to those facilities necessary for allowing the other Contractor to perform his obligations.

Although the clause does not cover payment, Clause 6.1 entitles the Contractor to be paid if he incurs additional cost and the provision of facilities could well be authorized by the Engineer under Clause 2.3(g). Such authorization would entitle the Contractor to additional payment. To avoid the possibility of claims for additional payment the Engineer should detail in the Specification any facilities required to be supplied by the Contractor for other contractors so that the cost of such provision can then be deemed to be included in the tender price.

Comparison with ICE 5

The first and fundamental obligation laid upon the Contractor as set out in Clause 3.1 is to perform and complete the works. Indeed, this must be so, as it is precisely what the Employer is paying for and would arise in any event as a necessary implication of the law. The analogous Clause 8(1) in ICE 5 is slightly wider in that it requires the Contractor to 'construct, complete and maintain' the Works whereas Clause 3.1 only requires him to 'perform and complete' them. ICE 5's requirement to 'maintain' the works has not been omitted, as it appears in a slightly different form in Clause 4.7 (Rectification of defects) of the minor works form.

Clause 3.1 further requires the Contractor to provide all necessary supervision, labour, materials, plant, transport and temporary works, thereby combining the rest of ICE 5 Clause 8(1) with the substance of Clause 15(1). The requirement of Clause 3.2 requires the Contractor to take care of the Works from their commencement until 14 days after they are completed and thus it echoes the requirements of Clause 20(1) of ICE 5. By 'completion' is meant the issue by the Engineer under Clause 4.5 of a Certificate of Practical Completion for the whole of the Works. As in ICE 5, the 14 day overrun of the Contractor's responsibility is intended to give the Employer time to take out such insurances.

The minor works form permits the Employer to take over parts of the Works prior to completion of the whole of the Works. The equivalent Clause 48 in ICE 5 refers to such parts as are specified in the Appendix as 'sections' and ICE 5 permits a further sub-category, 'a part', to be taken over by the Employer when the Works have both passed any final test and been completed to the satisfaction of the Engineer. There is no such provision for the Engineer to take over a part in the minor works form (although the equivalent to a section in ICE 5 is called a part in the minor works form).

In both Contracts, the Contractor has to take full responsibility for outstanding work which is undertaken after the issue of a Completion Certificate. In both Contracts, the Contractor has to make good 'any damage, loss or injury which happen to the Works or any part thereof' and do so at his own expense. Such damage however, is covered by the insurance of the Works in both Contracts. The two Contracts have been designed so that the policies of insurance for one will cover the other. (See Clause 20(2) of ICE 5.)

The Excepted Risks are the same in both contracts. This is to ensure that Contractors who take out annual all-risks insurance policies can undertake a mixture of contracts under ICE 5 and the minor works form without worrying about the validity of their insurance cover (Clause 1.5 of the minor works form and 20(3) of ICE 5).

Clause 3.4 requires the Contractor to notify the Engineer of the person duly authorized to receive instructions on behalf of the Contractor. The equivalent Clause 15(2) of ICE 5 is much more stringent. This requires the Engineer to give approval to a representative — which approval can at any time be withdrawn. Secondly, such person is to give his full time to superintendence of the Works and finally, such person is to be responsible for the safety of all operations.

Clause 3.5 requires the Contractor to take full responsibility for the setting out of the Works and the adequacy, stability and safety of site operations and methods of construction. This is an amalgamation of the requirements of Clause 8(2) and 17 of ICE 5. While the legal effect is the same in both sets of conditions, the resulting responsibility placed upon the Contractor may in practice be heavier under the minor works form in that there will seldom be a continuing presence on the Site by the Engineer, at least at the level of expertise which is normally expected for projects subject to ICE 5. It is thus, if anything, somewhat more important under the minor works form for the Contractor to keep himself informed of events on the Site, as he cannot rely on the Engineer's staff to pick up anything which his own staff have missed.

Clause 3.6 makes the Engineer responsible for the provision of any necessary instructions or drawings or other information. This reflects the requirements of Clause 7(1) of ICE 5, except that the Contractor is not made responsible for requesting information necessary to complete the Works. This throws a greater burden onto the Engineer to ensure that all the necessary information is given in good time to the Contractor or else the Contractor will have a claim under Clause 6.1.

The Contractor under Clause 3.7 is not responsible for the design of the Works except where expressly stated in the Contract, nor for the design of any temporary works designed by the Engineer. This is identical to the requirements of Clause 8(2) of ICE 5.

The provision for encountering artificial obstructions or physical conditions is the same in both Contracts, except in one respect. The exception is that under Clause 3.8 of the minor works form the Contractor *must* inform the Engineer of each and every condition or obstruction which he thinks could not have reasonably been foreseen by an experienced contractor, whereas under ICE 5 he is only bound to do so if he intends to make a claim for extra cost, and/or an extension of time.

It would seem that this extra complication could result in a flood of minimally necessary documentation, but it must be remembered that on an ICE 5 project there will almost always be a continuous presence of suitably qualified representatives of the Engineer on the Site, whereas on a Contract under the minor works form that will not be so. The Engineer must therefore rely upon the Contractor to keep him informed of all relevant site conditions.

Under ICE 5 the Contractor is required to inform the Engineer of the measures he is taking, or proposing to take. There is no such obligation under the minor works form.

Under ICE 5 Clause 12(2) the Engineer can

(i) ask the Contractor for an estimate of the costs of the measures he is taking or proposing to take,

(ii) approve in writing such measures, with or without modification,

(iii) give written instructions as to how the physical conditions or artificial obstructions are to be dealt with, or

(iv) order a suspension under Clause 40 or a variation under Clause 51.

Under the minor works form, however, the Engineer has all such powers with the exception of requesting an estimate. He can give instructions for any measures necessary to overcome or deal with any obstruction or condition falling within Clause 3.8 under Clause 2.3(e), order a suspension of the Works under Clause 2.3(c), or order a variation under Clause 2.3(a). In both Contracts the methods of payment are the same.

Whilst the express power to order an extension of time is not given in Clause 3.8, it is specifically mentioned as a ground for an extension under Clause 4.4(c). The risk, therefore, from the point of view of the Employer on the Contractor finding unforeseen physical conditions or artificial obstructions is roughly the same under both contracts.

Clause 3.9 requires the Contractor to afford reasonable facilities to other Contractors employed by the Employer on the Site. This echoes the requirements of Clause 31(1) of ICE 5, but is somewhat more limited. ICE 5 can cover other Contractors working near the Site, whereas the requirements of the minor works form are only for those who will work upon the Site and it would be implied as well that their work must be in relation to the Contract Works.

The Minor Works Clause does not specifically mention payment for these facilities, as does Clause 31(2). The Contractor nevertheless would get paid because any instruction to give facilities to contractors would be deemed to be issued under Clause 2.3(a) and qualify for payment under Clause 6.1.

Comparison with ICE 6

Save that under Clause 12 the Contractor now has a duty to inform the Engineer of any condition or obstruction which in his opinion could not have been foreseen whether or not he intends to make a claim or seek instructions in connection therewith (thereby adopting the position under Clause 3.8 of the minor works form), the provisions of ICE 6 are in substance broadly the same as those of ICE 5.

Legal comment

Position of the Contractor

Under the general law, and in the absence of a contract, the Contractor will be deemed to be under an obligation to carry out the Works using due skill and care and to complete them within a time which is reasonable in all the circumstances. In addition, the Supply of Goods and Services Act 1982 imposes a specific duty *inter alia* to

(i) use reasonable skill and care when undertaking any design;

(ii) ensure that materials supplied are of marketable quality and reasonably fit for their intended use;

(iii) give a warranty that the supplier has good title to any goods and materials supplied; and

(iv) complete the work within a reasonable period.

The Act thus gives statutory expression to the common law requirements and takes some of them further. It follows that any contract which purports to cut down the statutory provisions may be subject to revision by the court should a dispute ever get that far.

The minor works form is essentially a contract for services, and not one for the supply of goods. Thus, while materials should be of marketable quality and reasonably fit for their purpose (and the Contractor should have good title to such materials), it does not follow that the facility or structure of which those materials will in due course form a part must also be 'fit for its purpose'.

Moreover, the requirement that the Works be completed within a reasonable period is overridden by the express minor works requirement that they be completed by the specified completion date. To the extent that the Contractor may be required to carry out design (normally of temporary works and, exceptionally, specified parts of the permanent works), it has already been pointed out that the standard of performance is the use of due skill and care.

That having been said, the functions and obligations of the Contractor under the minor works form are largely the same as under ICE 5, and practitioners familiar with the latter should find little difficulty in following the former. There are , of course, some matters of detail where the two forms differ slightly, but the general tenor of the two documents is the same.

Unforeseen conditions and obstructions

This subject is discussed at length in Chapter 18. However, it is convenient to deal here with the policy behind the retention of this provision in the minor works form.

During the extensive consultations which preceded publication of the minor works form it was argued with some force that, as the new form was intended for use only on smaller, low-risk projects, there was no need for the wide-ranging provisions of ICE 5, Clause 12, which relieve the Contractor of the risk of encountering adverse physical conditions or artificial obstructions. But it was also argued with equal force that, precisely because the new form was intended for low-risk projects, unforeseen escalation of that risk should fall upon the Employer since, unlike the situation in the building industry, all civil engineering projects, regardless of size, were inherently vulnerable to the unexpected once the ground was opened.

In the event, it was decided that a Clause 12 provision was merited. Accordingly, as drafted, Clause 3.8 includes the substance of Clauses 12(1) to 12(4) of ICE 5 and, save for a welcome reduction in verbiage, is with one exception (the Contractor's duty to report in any event) of the same legal effect.

Chapter 7

STARTING AND COMPLETION

Introduction

The second fundamental obligation of the Contractor is to complete the Works within the period stipulated in the Contract. The Employer has a very real interest in the time that the Works are completed, for it is from that moment that he can take possession of the Works, either for the purposes of their completion by others or for their use. In either case, he is liable to suffer loss if the Works are not completed on time.

The starting date is either stipulated in the Contract or given to the Contractor by the Engineer, and the Contractor must complete the Works within a period specified in the Contract. Provision may be made for parts of the Works to be completed in times shorter than the period for the completion of the whole of the Works. If the Contractor fails to complete within the specified periods, he is liable to pay liquidated damages for lateness.

Provision is made for the periods for completion to be extended as a result of events for which the Employer takes responsibility.

Starting date

> 4.1. The starting date shall be the date specified in the Appendix or if no date is specified a date to be notified by the Engineer in writing being within a reasonable time and in any event within 28 days after the date of acceptance of the Tender. The Contractor shall begin the Works at or as soon as reasonably possible after after the starting date.

Provision is made in Item 5 of the Appendix to the Conditions of Contract for the starting date to be written into the Contract. This should always be done if the date is known when tenders are invited but, if it is not, the Engineer must notify the Contractor in writing of the starting date as soon as it is known.

Such notification must be made within a reasonable time of the acceptance of the Contractor's tender, with a maximum of 28 days from such acceptance. As the Notes for guidance indicate (Note 8) acceptance should follow within two months of the date for submission of tenders, the starting date should therefore normally be within three months of the submission of tenders, and it is on this assumption that the Contractor will have based his prices.

What is a reasonable time will depend on the circumstances of each project (with the proviso that anything more than 28 days is, by definition, unreasonable). However, most Contractors will normally need about a fortnight to organize their administration, prepare the Site and call up materials and plant. As the Contractor is required to begin the Works at or as soon as reasonably possible after the starting date, it seems reasonable to expect the Engineer to notify the starting date about 14 days in advance of that date and no more than 14 days after the acceptance of tender.

Should the Engineer fail to give notification within the 28 days prescribed by Clause 4.1, it is arguable that he will thereby put the Employer into breach of contract. However, such breach is unlikely to lead to a claim for delay or for increased prices unless the delay in notification is significant either in the light of the overall time for completion or because of some intervening and inflationary event (such as a major increase in oil prices). In an extreme case such delay might give rise to a repudiatory breach entitling the Contractor to treat the Contract as having been abandoned by the Employer, but the circumstances which would justify such a result would be very rare. Nevertheless, a delay which (for instance) resulted in summer working having to be performed in winter would clearly give rise to a legitimate claim for extra cost and possibly for an extension of time as well.

Provision of Site

The Employer is responsible for the provision of both the Site and access to it, and it is important that these should be available prior to the starting date notified by the Engineer under Clause 4.1.

> NG9. Access as necessary to the Site should be available at the starting date under Clause 4.1.

The reason for this is that under Clause 4.1 the Contractor must begin the Works at or as soon as is reasonably possible after the starting date. Clearly, if the acceptance and starting date were contained in the same letter, it would seldom therefore be possible for the Contractor to start the following day. There must be sufficient flexibility to enable the Contractor to do any necessary planning and ordering or obtaining materials and plant so that, when the actual work does commence, it can continue uninterrupted.

Time for completion

> 4.2. The period or periods for completion shall be as stated in the Appendix or such extended time as may be granted under Clause 4.4 and shall commence on the starting date.

The period for completion of the whole of the Works is to be found in Item 6 of the Appendix to the Conditions of Contract. Where parts of the Works are to be handed over earlier, then these parts have to be defined in Item 7 of the Appendix. It is not possible for the Engineer to insist on early completion of any sections of the work that are not defined as parts in the Appendix.

> NG13(2). *Clause 4.2 (Period for completion).* If the Contract requires completion of parts of the Works by specified dates or within specified times, such date or times and details of the work involved in each part must be entered in the Appendix to the Conditions of Contract.

It will be seen that the space left for description of the parts in the Appendix is very small, and a detailed description of any parts so designated should be given in the Specification. Periods for completion should always be stated in weeks. The reference to dates is for contracts where a starting date has been specified in Item 5 of the Appendix to the Form of Tender.

Programme

4.3. The Contractor shall within 14 days after the starting date if so required provide a programme of his intended activities. The Contractor shall at all times proceed with the Works with due expedition and reasonably in accordance with his programme or any modification thereof which he may provide or which the Engineer may request.

The Contractor is only required to submit a programme 'if so required'. Any such request should therefore be made by the Engineer sufficiently in advance of the date on which it is needed to allow the Contractor a reasonable time to prepare it, since on a small project the Contractor's estimators may not have needed to prepare one in order to complete the tender. Ideally, a programme should be requested in the tender documents (to be delivered either with the tender or within a specified time of acceptance), or at the latest within the Engineer's notification of the starting date.

Since the design of the Works under the minor works form is to be complete in all essentials before tenders are invited (Note 1(f) in Notes for guidance), the main purpose of the programme is to enable the Engineer to organize his supervision of the Works. It will also enable him to check the progress of the Works as they are executed to ensure that, so far as is possible, the Works will be completed on time.

While there is no specific provision in the minor works form to that effect, the fact that a programme can be requested necessarily implies both that the Engineer can request that the programme as submitted be modified and that he can ask for further information in support of the programme. There is no requirement that any such programme shall be approved or consented to by the Engineer, but he will obviously tell the Contractor of anything about it which he considers inappropriate and, in an extreme case, he could issue appropriate instructions under Clause 2.3(g).

The Contractor is required 'at all times [to] proceed with the Works with due expedition and reasonably in accordance with his programme or any modification thereof which he may provide or which the Engineer may request' and failure to do so is *prima facie* a breach of contract.

The requirement to proceed 'with due expedition' is important for both Employer and Contractor, and this or a similar term is found in most building and civil engineering forms of contract. If the Contractor consistently fails to make reasonable progress, has insufficient resources on site or takes labour, materials or plant off site for no good reason, the Engineer may be justified in notifying him that he is in breach of contract for not proceeding with due expedition.

If after due warning the Contractor still persists in failing to make progress, his conduct could be deemed to amount to repudiation of the Contract. In such a case, and as there is no forfeiture clause in the minor works form, the Employer could choose either to wait until the contractual date for completion and then terminate the Contract for failure to complete on time or to exercise his common law rights to accept the Contractor's repudiation, expel him from the Site and employ others to complete the Works. However, on a short contract it will normally be preferable to wait for the completion date, since common law forfeiture will almost certainly lead to litigation.

As for the requirement to proceed 'reasonably in accordance with [the] programme', this can only operate if a programme has been requested and/or submitted. If there is such a programme, failure reasonably to comply with it may also, at least in theory, be a breach of contract. However, it is probably better to see this part of the requirement as a yardstick against which 'due expedition' is to be measured. After all, if the Contractor is falling seriously behind in his programme, the Engineer's obvious response is to request a modification of the programme aimed at completion on time.

Extension of time

The purpose of an extension of time is to enable the contract period to be adjusted when circumstances arise which are the responsibility of the Employer, and which would prevent the Contractor from fulfilling his obligation to complete within the contract period. An extension of time also preserves the Employer's right to liquidated damages which might otherwise be lost.

Most construction contracts include an extension of time clause, and so does the minor works form.

> 4.4. If the progress of the Works or any part thereof shall be delayed for any of the following reasons:
> (a) an instruction given under Clause 2.3 (a) (c) or (d);
> (b) an instruction given under Clause 2.3 (b) where the test or investigation fails to disclose non-compliance with the Contract;
> (c) encountering an obstruction or condition falling within Clause 3.8 and/or an instruction given under Clause 2.3 (e);
> (d) delay in receipt by the Contractor of necessary instructions, drawings or other information;
> (e) failure by the Employer to give adequate access to the Works or possession of land required to perform the Works;
> (f) delay in receipt by the Contractor of materials to be provided by the Employer under the Contract;
> (g) exceptional adverse weather;
> (h) other special circumstances of any kind whatsoever outside the control of the Contractor
>
> then provided that the Contractor has taken all reasonable steps to avoid or minimize the delay the Engineer shall upon a written request by the Contractor promptly by notice in writing grant such extension of the period for completion of the whole or part of the Works as may in his opinion be reasonable. The extended period or periods for completion shall be subject to regular review provided that no such review shall result in a decrease in any extension of time already granted by the Engineer.

The clause only applies to delays suffered during 'the progress of the Works or any part thereof'.

Reasons for extensions of time

> (a) an instruction given under Clause 2.3 (a) (c) or (d);

These sub-clauses to Clause 2.3 cover respectively, variations, suspensions and changes in sequence of the Works.

Commentary

1. *Variation orders*

2.3(a) any variation to the Works including any addition thereto or omission therefrom;

It is only the delay necessarily caused by the Engineer's variation order which will entitle the Contractor to an extension of time. However, if such delay has occurred, the only limit which the Contract places on the resulting grant of an extension is that it shall be reasonable.

This, in turn, means 'reasonable in all the circumstances'. The Contractor is required by the proviso at the end of the clause to take all reasonable steps to avoid or minimize delay and, if he is successful in avoiding *all* delay he will *not* be entitled to an extension (although he may be able to claim additional costs for disruption). But, provided that he does so minimize the delay, that which remains will entitle the Contractor to an appropriate extension of time.

2. *Suspension orders*

2.3(c) the suspension of the Works or any part of the Works in accordance with Clause 2.6;

The Engineer has power to suspend the Works on the following grounds:

(i) for the proper execution of the work;
(ii) for the safety of the Works or any part thereof; or
(iii) by reason of weather conditions.

If the Engineer suspends the whole of the Works for reasons other than a default of the Contractor, then in due course the Contractor will in principle become entitled to an extension of time equal to the period of suspension. However, such extension could be longer if the effect of the suspension was, for instance, to result in summer working having to be performed in winter. Similarly, if there are steps which the Contractor can and should take to minimize the delay, the extension will be reduced accordingly.

If only part of the Works is suspended, the same principles apply. However, it will be for the Contractor to demonstrate (if that part of the Works is not the subject of a separate completion date under the Contract) that, but for the suspension, not only would that part have been completed by the date he intended, but also that the completion of the whole of the Works was also affected.

3. *Changes in sequence*

2.3(d) any change in the intended sequence of the Works;

The intended sequence is clearly that either stipulated by the Contract or intended by the Contractor. In the latter case it will, of course, be for the Contractor to prove to the Engineer's satisfaction what his intentions were and, if a programme has not been submitted, this could be difficult. Be that as it may, if the Engineer gives an instruction which changes the intended sequence, and that instruction gives rise to delay, then, provided that the Contractor has taken all reasonable steps to minimize the delay, he will be entitled to an appropriate extension of time.

The time at which the instruction is given will have a material effect upon the delay. If it is given well in advance of the programmed date for any of the re-timed work, then clearly the Contractor should be able to avoid any delaying effects of the change.

4. *Tests and investigations*

4.4(b). An instruction given under Clause 2.3(b) where the test or investigation fails to disclose non-compliance with the Contract.

2.3(b). carrying out any test or investigation;

Such tests may include acceptance tests of workmanship or materials, testing the acceptability of the formation or ground prior to the placing of foundations or other tests or investigations in connection with the design of the Works. Provided such tests delay the Contractor, he will be entitled to an extension of time unless the tests disclose work which is not in accordance with the requirements of the Contract.

However, Sub-clause 4.4(b) only deals with delays caused by the tests themselves; the consequences, such as design changes, will be dealt with as a result of other instructions issued by the Engineer.

5. *Unforeseen conditions or obstructions*

4.4(c). encountering an obstruction or condition falling within Clause 3.8 and/or an instruction given under Clause 2.3(e);

2.3(e). measures necessary to overcome or deal with any obstruction or condition falling within Clause 3.8;

If the Engineer is satisfied that the artificial obstruction or physical condition could not reasonably have been foreseen by an experienced contractor, the Contractor is entitled to an extension of time. His entitlement may be reduced if he fails to give written notice of the obstruction or condition to the Engineer as early as practicable. The amount by which the extension would then be reduced would be the difference between the date on which the Contractor actually gave notice or the Engineer became aware of the conditional instructions and the date upon which it was practicable for him to give written notice.

A commentary on Clause 3.8 is given in Chapter 6, and 'the unforeseen' is discussed in Chapter 18. Extensions of time can arise from encountering unforeseen obstructions under two headings. Firstly, the time that the Contractor necessarily takes in dealing with and overcoming the condition or obstruction, and secondly, the additional time lost as a result of any instruction given by the Engineer under Clause 2.3(e).

As before, the Contractor has a duty to minimize delay and his entitlement can be reduced if he takes an unduly long time to decide what measures are needed to overcome the condition or obstruction or to begin taking such measures.

It should be noted that it is entirely possible that encountering an artificial obstruction or physical condition may entitle the Contractor to payment but not to extension of time because it has no effect upon completion of the whole of the Works, only upon the completion of a part of the Works. Similarly an entitlement to extension of time will not necessarily be accompanied by an entitlement to extra payment. These two aspects are quite distinct and must be dealt with separately and on their individual merits.

6. *Delayed instructions*

4.4(d). delay in receipt by the Contractor of necessary instructions, drawings or other information;

Under Clause 3.6 the Engineer is responsible for the provision of any necessary instructions, drawings or other information. As has already been stated, this should all have been accomplished preferably at tender stage and at any event before the starting date. Nevertheless, if subsequently the need arises for further information, this should be produced and given to the Contractor in good time so as not to disrupt or delay the due progress of the Contractor's operations.

If, however, such information is delivered late and the Contractor suffers delay as a result, he may well be entitled to an extension of time on that account. Moreover, there is no obligation upon the Contractor to ask for such information, since it is the Engineer's duty to volunteer it in good time. Nevertheless, should the Contractor realize that certain information is lacking, he would be prudent to draw the Engineer's attention to the fact as soon as possible.

7. *Possession or access to Site*

4.4(e). failure by the Employer to give adequate access to the Works or possession of land required to perform the Works;

The provision of both adequate access to and possession of the Site is the responsibility of the Employer. Where access is restricted by width, height, load capacity or the time when it is available then the Engineer should check the Contractor's requirements prior to the starting date.

8. *Employer's materials*

4.4(f). delay in receipt by the Contractor of materials to be provided by the Employer under the Contract;

Where materials are to be supplied by the Employer, the Contract Documents should state when they will be made available. If such information cannot be included in the Contract the Engineer should check with the Contractor before the starting date when the Contractor expects those materials to appear on the Site, and then satisfy himself that those delivery dates can be met. If such enquiries are not made at the proper time and the Contractor's expectations with regard to delivery are not realized, an extension of time may be necessary, quite apart from a claim for any costs occasioned by the delay.

9. *Weather conditions*

4.4(g). exceptional adverse weather;

Although weather conditions as such are a Contractor's risk, this sub-clause permits an extension of time where the weather is 'exceptionally adverse', even though additional costs arising from such weather cannot be recovered. Whether or not conditions are in fact exceptionally adverse must be judged in the light of what is normal for the place and the time of year at which they occur. The type of weather is not limited in any way, since drought, deluge, temperature and wind conditions can all have unduly severe effects on operations which are sensitive to such parameters.

10. Other special circumstances

4.4(h). other special circumstances of any kind whatsoever outside the control of the Contractor;

The scope of this provision is very wide indeed, and is capable of accommodating almost any event which might fairly and reasonably entitle the Contractor to an extension. Thus delays arising from the Excepted Risks (Clauses 1.5 and 10.1) or the provision of facilities for other contractors (Clause 3.9), neither of which is specifically mentioned in this part of the Contract, can conveniently be dealt with under this sub-clause.

Similarly, any other cause of delay which can be construed as being at the Employer's risk and is not otherwise covered can be included here, as can matters wholly beyond the control of either party, such as additional public holidays, changes in the law or any relevant regulations or by-laws, or even the effects of strikes and other industrial action other than by the Contractor's own staff and operatives.

Action by Engineer

4.4. . . . then provided that the Contractor has taken all reasonable steps to avoid or minimize the delay the Engineer shall upon a written request by the Contractor promptly by notice in writing grant such extension of the period for completion of the whole or part of the Works as may in his opinion be reasonable. The extended period or periods for completion shall be subject to regular review provided that no such review shall result in a decrease in any extension of time already granted by the Engineer.

The Contractor's duty to mitigate the Employer's loss is satisfied provided he takes 'all reasonable steps' to avoid or minimize the delay. There will usually be a careful balance to be struck between the duty to reduce time and the need to minimize expenditure resulting from the delaying event. When assessing whether the Contractor has taken all reasonable steps it will be necessary to examine the alternatives which might, if implemented, have led to excessive expenditure, as well as those which might have shown a reduction in cost — but bearing in mind the pitfalls inherent in an excessive reliance on hindsight.

The Engineer is required following a written request by the Contractor promptly to grant by notice in writing any extension of time to which in his opinion the Contractor is entitled. The dictionary definition of prompt is 'acting with alacrity' and alacrity means 'briskness, cheerful readiness'. Since one of the objects of an extension of time is to protect the Employer's right to liquidated damages, it is necessary for the Engineer to act promptly to enable the Contractor to take what measures he can to avoid incurring liquidated damages. This means that, as soon as the delaying event is identified, the Engineer should be prepared to grant on an interim basis such partial extensions of time as can readily be justified at that stage, and thereafter should keep such decisions under regular review.

It is of particular importance that the Contractor shall know what his entitlement to extensions of time may be, lest, in the absence of such assurance, he feels obliged to increase the resources committed to the project, which increases in due course may later be seen to have been unnecessary. In such circumstances the Contractor may well submit a claim for the costs of such unnecessary acceleration on the grounds of an implied breach of duty by the Engineer. It is a matter for conjecture whether such a claim would succeed, but it would inevitably lead to bad feeling and almost certainly to a messy arbitration.

From the practical point of view extensions of time should be examined at every progress meeting so that necessary facts can be obtained while they are still fresh and any further grant which may be justified can be made within a few days.

The only criterion under the Contract for granting extensions of time is that which the Engineer thinks is fair and reasonable in all the circumstances. It should, of course, go without saying that the Engineer's opinion must be objective and uninfluenced by what either the Employer or the Contractor would like to see done. That being so, the prudent Engineer will be cautious in considering what extensions to grant, since an over-generous grant cannot later be reduced or withdrawn, yet granting too little too slowly can prejudice the Contractor and put the Employer at risk. What certainly must *not* be done is for the Engineer to abdicate his responsibility and leave it all to be sorted out at the final account stage.

Finally, although an extension once granted cannot be reduced or withdrawn, the Engineer will be entitled when considering subsequent grants to take into account any factors, such as variations omitting part of the Works, early delivery of Employer's materials, early access to the Site, unusually advantageous weather and the like, which tend to reduce the Contractor's need for an extension. There would seem to be no reason why such favourable factors should not be offset against the effects of delaying events entitling the Contractor to subsequent extensions.

Completion

4.5(1). Practical completion of the whole of the Works shall occur when the Works reach a state when notwithstanding any defect or outstanding items therein they are taken or are fit to be taken into use or possession by the Employer.

The period for completion of the Works, as stated in Item 6 of the Appendix to the Conditions of Contract and referred to in Clause 4.2, ends when the Works reach practical completion. Practical completion does not mean total completion but is achieved when the Works reach a state in which they can be used by the Employer. Under Clause 4.7, the Contractor is then required to complete any outstanding items and make good any defects promptly after the Works have reached practical completion.

The requirement to reach practical completion rather than total completion is to the benefit of Employer, Engineer and Contractor alike. The Employer gains by obtaining access to the Works at the earliest possible date, the Engineer benefits from not having to do a minute survey prior to certifying that the Works have reached practical completion (because to certify total completion would mean that the Contractor has no further work to undertake). The Contractor benefits by the commencement of the Defects Correction Period as early as possible, thus reducing the period of his liability, obtaining release of the first half of the retention money, and assisting his cash flow.

4.5(2). Similarly practical completion of part of the Works may also occur but only if it is fit for such part to be taken into use or possession independently of the remainder.

There is no definition of a 'part', although there is provision for the early completion of parts designated in Item 7 of the Appendix to the Conditions of Contract.

It would seem that there is no restriction upon practical completion occurring for parts not so designated, if the Employer wishes to take possession of an undefined part. There is, however, no way in which the Engineer can instruct a part to be handed over by a specific date earlier than the date for the completion of the whole of the Works unless such a part is designated in the Appendix to the Conditions of Contract.

Completion certificate

4.5(3). The Engineer shall upon the Contractor's request promptly certify in writing the date upon which the Works or any part thereof has reached practical completion or otherwise advise the Contractor in writing of the work necessary to achieve such completion.

The Contractor's obligation in respect of time ceases on the date (as certified by the Engineer) upon which the Works have reached practical completion.

It is for the Contractor to start the process of certification. As soon as he considers that the Works have reached a stage where they are fit to be taken by the Employer into possession and use, the Contractor may submit a request in writing to the Engineer for the issue of a Certificate of Practical Completion. Upon receipt of such a request the Engineer must promptly either

(i) issue a certificate stating the date upon which in his opinion the Works reached practical completion (which may or may not be the date applied for by the Contractor), or
(ii) issue to the Contractor a list of all the items which must be carried out or rectified before he will issue such a certificate.

In the latter event the Contractor will automatically be entitled to a certificate once all the items on the Engineer's list have been satisfactorily dealt with. There is no provision for the Engineer to add further items to the list once the original list has been issued.

The effect of issuing a Certificate of Practical Completion for the whole of the Works is that:

(i) the Defects Correction Period referred to in Clause 5.1 will commence on the date stated in the certificate;
(ii) responsibility for the care of the Works will pass from the Contractor to the Employer 14 days after the date of issue of the certificate (Clause 3.2); and
(iii) one half of the retention money under Clause 7.4 will be released 14 days after the date of issue of the certificate.

The effect of issuing a certificate for only part of the Works differs in that only responsibility for that part of the Works will pass to the Employer (as in (ii) above), since neither the commencement of the Defects Correction Period nor the release of retention money is affected by completion of a part. However, there will be a corresponding reduction in the amount of liquidated damages payable in respect of those parts of the Works which have yet to be completed (Clause 4.6).

Finally, the minor works form contains no definition of a 'part' of the Works. It follows that any part can be the subject of a Certificate of Practical Completion provided only that it is fit to be taken into the Employer's use or possession independently of the remainder (Clause 4.5(2)). And the fact of possession or use by the Employer is itself enough to initiate the process of certification even though a certificate would otherwise be premature.

Liquidated damages

4.6. If by the end of the period or extended period or periods for completion the Works have not reached practical completion the Contractor shall be liable to the Employer in the sum stated in the Appendix as liquidated damages for every week (or pro rata for part of a week) during which the Works so remain uncompleted up to the limit stated in the Appendix. Similarly where part or parts of the Works so remain uncompleted the Contractor shall be liable to the Employer in the sum stated in the Appendix reduced in proportion to the value of those parts which have been certified as completed provided that the said limit shall not be reduced.

The liquidated damages payable by the Contractor in the event of not completing the Works in the period for completion are shown in Item 8 of the Appendix to the Conditions of Contract. Item 9 of the Appendix gives the limit of liquidated damages that the Contractor is liable to pay and once this figure is reached the Contractor is not liable in damages for further delay. Note for Guidance 13(5) gives guidance on the assessment of these damages.

NG13(5) *Clause 7.3 (Rate of retention and limit of retention).* The rate of retention should normally be 5%. A limit of retention has been inserted in the Appendix and this should normally be between the limits of $2\frac{1}{2}$% and 5% of the estimated final Contract value.

There is no provision for independent liquidated damages for late completion of the parts specified in the Appendix. When a part so specified or a part as defined by the Engineer reaches practical completion and the Engineer issues a certificate to that effect then the liquidated damages are reduced proportionally to the value of the Works handed over.

However, failure to complete a part within the period stated might entitle the Employer to damages at common law, but any damage would have to be both incurred and proved.

Unfortunately the drafting of this clause defining reduction in liquidated damages when a part is handed over is imprecise. However, since liquidated damages can be adjusted in the final account, it is suggested that the eventual reduction should be calculated in the proportion that the value of the Works handed over bears to the final contract sum. If reductions have to be made on an interim basis then it would be sensible to use the tender sum instead of the final contract sum at that stage.

Rectification of defects

4.7. The Contractor shall rectify any defects and complete any outstanding items in the Works or any part thereof which reach practical completion promptly thereafter or in such manner and/or time as may be agreed or otherwise accepted by the Engineer. The Contractor shall maintain any parts which reach practical completion in the condition required by the Contract until practical completion of the whole of the Works fair wear and tear excepted.

The Contractor's obligation is promptly to complete works outstanding at the date of practical completion and also to repair any defects which appear during the Defects Correction Period, normally as soon as they arise.

After practical completion the Employer will normally have taken the Works into possession and/or use, and the Contractor may well have no personnel remaining on the Site. It is thus clearly in the mutual interest of both parties that sensible arrangements and timings are agreed for the completion of outstanding work or the rectification of defects so as to minimize any resulting inconvenience to Employer and Contractor alike. Such arrangements will normally be settled between the Contractor and the Engineer and, once agreed, progress will be checked by, and the work examined and inspected by, the Engineer.

Comparison with ICE 5

Minor works form Clauses 4.1, 4.2 and 4.3 between them cover the same ground as Clauses 41 and 43, and part of Clause 14 of ICE 5.

Under Clause 4.1, the starting date (ICE 5 Date of Commencement of the Works) is to be either that specified in the Appendix to the Conditions of Contract, or notified by the Engineer in writing, in which case it must be within a reasonable time, in any event not more than 28 days after the tender is accepted. In this it differs from the ICE 5 provision, which has no specified date, and no prescribed limit to the time that the Engineer may take to issue the notice of commencement. The only stipulation is that it shall be issued within a reasonable time of the award of the Contract. Again no guidance is given as to what is reasonable.

It follows that tenders for minor works ought not to be accepted until it is known when the Works may proceed. If the starting date is uncertain, there is more risk for the Employer under the minor works form, because delay beyond the 28 days could lead to damages for repudiation or delay.

However, should circumstances arise after the tender has been accepted in which, for good reason, the Employer needs to postpone the start of the Works substantially, the problem ought not to be met by delaying the Engineer's notification (which can and does happen under ICE 5). Instead, either a revised starting date should be agreed with the Contractor (in which case he may well bargain for a suitable adjustment to his tender price) or the Contract should be terminated.

Under both forms of contract, the period or periods for completion of the Works must be stated in the Appendix (either to the Conditions of Contract to the minor works form, or to the Form of Tender under ICE 5), and Clauses 4.2 and Clause 43 respectively are thus of identical effect. Both clauses include a reference to the provisions for extensions of time.

The express requirement in ICE 5, Clause 41, that the Contractor should proceed with the Works with due expedition and without delay *in accordance with the Contract* appears, slightly modified, in minor works Clause 4.3, where the obligation is to proceed with due expedition and *reasonably in accordance with his* (the Contractor's) *programme*. While the minor works provision appears at first sight to be milder than that in ICE 5, common sense would suggest that the minor works provision probably expresses what usually happens in practice under ICE 5.

On the other hand, both expressly require the Contractor to begin the Works 'at or as soon as reasonably possible after the starting date', or 'commencement date' as the case may be.

Success in starting and finishing the Works on time logically presupposes forward planning, and the minor works form follows this logic by including the submission of the Contractor's programme at Clause 4.3. This clause differs from ICE 5, Clause 14(1) in that the submission is to be made within 14 days after the starting date, which itself (under Clause 4.1) may be up to 28 days after the acceptance of tender, whereas the ICE 5 requirement is for submission within 21 days of acceptance of the tender. The Engineer could well receive the programme at a later date under the minor works form.

The difference may, perhaps, be explained by assuming a need under ICE 5 for the programme to be submitted before construction starts — a need which, arguably, will not usually exist for the simpler kinds of project likely to be carried out under the minor works form. The same explanation may also serve to cover the fact that under the minor works form the Contractor need only submit his programme if asked to do so, whereas under Clause 14(1) it must be submitted in any event.

A further difference is the omission from Clause 4.3 of the requirement in the second sentence of ICE 5, Clause 14(1), that the Contractor 'shall at the same time [as he submits his programme] also provide in writing . . . a general description of the arrangements and methods of construction' which he proposes to adopt.

No such expressed provision appears in the minor works form, but, should the Engineer feel that such information ought to be submitted, his power under Clause 2.3(g) to give instructions for 'the elucidation or explanation of any matter to enable the Contractor to meet his obligations' might suffice to secure an appropriate submission.

On the other hand, the primary purpose of Clause 2.3(g) is clearly to enable the Engineer to explain the Works to the Contractor, and not necessarily vice versa, so its usefulness for the purpose of securing a Contractor's method statement may be open to question. In serious cases, however, suspension of progress 'for the proper execution of the Works' or 'for the safety of the Works' can be ordered under Clause 2.6(1)(a) or (b).

Minor works Clause 4.4 encapsulates the whole of the provisions of ICE 5, Clause 44, and is essentially the same in its effect. The grounds for granting extensions of time are the same, and are, moreover, set out in a manner which is far easier to read and comprehend. The minor works provision, however, presupposes that the Contractor will first submit a request in writing for an extension, after which the Engineer must grant such extension 'as may in his opinion be reasonable', whereas under ICE 5 he must assess and grant the extension to which in his opinion the Contractor is 'fairly entitled'.

The test to be applied is thus somewhat less strict under the minor works form, but no doubt in practice Engineers will take much the same view of each particular situation, whichever conditions apply.

The three-tier consideration of extensions of time under Clause 44 — at the initial stage, at the date for completion of the Works, and at the date of issue of the Certificate of Completion — is condensed into the minor works provision for 'regular review'. The full ICE 5 procedure is clearly inappropriate for projects which for the most part will have both started and finished well within a calendar year.

The limitation in Clause 44(4) that no subsequent grant shall reduce any earlier grant, however, is retained as part of the regular review under Clause 4.4, and, in addition, the minor works form includes a proviso that the Contractor 'shall have taken all reasonable steps to avoid or minimize delay' as a condition precedent to any grant of extension of time, which proviso is absent under ICE 5. Any refusal to proceed before a grant is made (by no means unknown under ICE 5) will of itself disqualify the Contractor from receiving a grant under the minor works form. It should be noted that there is no minor works provision similar to ICE 5, Clause 45, on night and Sunday work, nor anything analogous to the so-called 'acceleration' provisions in ICE 5 Clause 46 (rate of progress).

However, for the short-duration projects for which the minor works form is intended, the need for such provision is arguably much less. Clause 45 matters can probably be handled adequately in the Specification, and Clause 46 matters by provision of adequate liquidated damages. Should any particular project appear to need Clause 45 or Clause 46 protection for the Employer (which should be a rare situation), consideration should be given to the use of the ICE 5 conditions.

Substantial completion, under ICE 5, is called practical completion under the minor works form. Minor works Clause 4.5 is otherwise a commendably concise summary of the somewhat complex provisions of ICE 5, Clause 48; in intention it is the same, and in substance it is, arguably, a great improvement. Thus Clauses 4.5 (1) and 4.5 (2), dealing with the whole and part of the Works respectively, define practical completion as occurring 'when the Works reach a state when notwithstanding any defect or outstanding items therein, they are taken or are fit to be taken into use or possession by the Employer'.

It follows that the actual use or possession by the Employer will entitle the Contractor to a Certificate of Practical Completion under Clause 4.5(3), whether or not such use or possession is premature, and whether or not the Works would, but for that use or possession, in fact be deemed to have reached a state which would justify a certificate. This is clearly the right approach. The legal effect of a Clause 4.5 certificate under the minor works form, as it should be, is to vest in the Employer an immediate right of possession of the Works, when he gains access thereto or uses them. This, in turn, is fundamental to the question of insurance and the timing of the transfer of risk from Contractor to Employer.

On the other hand, it may frequently suit the Employer's convenience to avoid taking possession, but still to have use of part of the Works, in order to avoid ensuing maintenance costs, or to delay the release of retention money, or for some other non-monetary reason. Although this should not happen, under ICE 5, Clause 48(2)(b), it is possible for the Engineer to argue that the Works have not been completed to his satisfaction when the Employer has taken possession. Thus, while the effect of Clause 48 of ICE 5 is similar to Clause 4.5 of the minor works form, it is perhaps easier to delay the issue of the Completion Certificate under ICE 5.

The 65 lines of text in ICE 5, Clause 47 (Liquidated Damages) are condensed in the minor works Clause 4.6 to just 10 lines. Moreover, the complex and, in practice, seldom fully understood provisions of Clause 47(2) for 'sectional' damages are wholly omitted, the same effect being amply produced in a single sentence in Clause 4.6.

In part, this is clearly because 'sectional' damages are far less likely to be needed in simple, quick projects. In those very few cases where failure to complete part of a minor works contract on time is likely to result in disproportionately serious injury to the Employer's interests, it should be quite simple to elaborate upon Clause 4.6 in the Specification, or even by suitably amending the Appendix to the Conditions of Contract. Indeed, Clause 47 of ICE 5 seems already to have given rise to some particularly knotty problems of legal interpretation, where the Appendix to the Form of Tender has been completed in an inconsistent manner, or even left partially completed.

Two points from ICE 5 have not been carried over into the minor works form. The first is the statement in Clause 47(3) that sums payable under the Clause are paid as liquidated damages and not as a penalty, but, as penalties are in any event unenforceable at law (at least in England and Wales), its omission makes no difference. The second is the express power under Clause 47(4) for the Employer to deduct liquidated damages, and the accompanying provision in Clause 47(5) for their reimbursement (the latter standing or falling with Clause 47(2) of course). Nothing in the minor works form expressly empowers the Employer to deduct, but the omission is hardly likely to cause difficulty in practice.

In this connection, however, it could be argued that under minor works form Clause 7.3, the Engineer should deduct liquidated damages from any certificate thereunder, whereas, under ICE 5, it is clearly for the Engineer to certify in full, leaving it to the Employer to deduct liquidated damages from the sum certified. It will be interesting to see what the courts will make of this difference, should the point ever be taken there.

Finally, there is provision in the minor works form for a limit to the amount of liquidated damages payable by the Contractor. There is no such limit in ICE 5.

Minor works Clause 4.7 (Rectification of defects) requires the Contractor to complete any items of work which remain outstanding at practical completion 'promptly thereafter or in such manner and/or time as may be agreed or otherwise accepted by the Engineer'. In addition, where parts of the Works reach practical completion early, he must maintain such parts 'in the condition required by the Contract' until practical completion of the whole of the Works (fair wear and tear excepted) but not, it seems, thereafter.

This differs from the generally assumed duty under ICE 5 to continue such maintenance until the end of the Period of Maintenance, but it may be that this general assumption is based on a misunderstanding of ICE 5, Clause 49(2), which, although less crisply drafted than the minor works form, does not expressly refer to 'duty to maintain'.

Minor works, Clause 4.7, also includes a general duty on the Contractor to rectify defects, and this is repeated in Clause 5.2 (Cost of remedying defects) for defects due to materials or workmanship not in accordance with the Contract, which appear during the Defects Correction Period.

The overall effect, therefore, between the two contracts, would appear to be similar, except that the minor works form has taken a two-stage approach.

Comparison with ICE 6

In drafting ICE 6 the opportunity seems to have been taken to bring the revised text more or less into line with the minor works form and away from ICE 5. Thus the Works Commencement Date is to be either the date specified in the Appendix to the Form of Tender or a date within 28 days of the award of the Contract or 'such other date as may be agreed between the parties', and Clause 48(3) of ICE 6 provides for 'premature use by the Employer', whilst the provisions for liquidated damages in ICE 6, although more complex than those in the minor works form, have also abandoned the confusing and potentially ambiguous three-column provisions of ICE 5. There are also some purely semantic revisions such as Defects Correction Period instead of ICE 5's Maintenance Period and Defects Correction Certificate in place of Maintenance Certificate.

ICE 6 now provides that when a part of the works is occupied or used by the Employer the Contractor will be entitled to a Certificate of Substantial Completion in respect of that part. The provision that it must also be complete to the satisfaction of the Engineer has been deleted. This brings ICE 6 into line with the minor works form.

Those matters apart, however, ICE 6 has in substance followed ICE 5 (albeit with fewer words, better layout and, hopefully, less room for potential ambiguity) and the above comparison of the minor works form with ICE 5 remains valid.

Legal comment

It is an express requirement of the minor works form (as it is in ICE 5, ICE 6 and many other standard forms of construction contract) that the Contractor will complete the Works on or before the completion date (or dates) stipulated in the Contract. To that extent time is of the essence, and a Contractor failing to complete on time will automatically be in breach of contract.

It may be, however, that through no fault of his own, the Contractor will be prevented from completing on time, either through some hindrance or prevention on the part of the Employer, his servants or agents (including the Engineer) or by some factor or event wholly outside the control of either party to the Contract. The Contract therefore empowers the Engineer to grant extensions of time, that is, in effect to amend the Contract by 'moving the goal posts' and adjusting the contractual date for completion to suit the new situation so far as it is fair and reasonable so to do.

There is also another aspect to this ability to grant extensions of time. If the Contractor is in breach with regard to time, the Employer is entitled to a remedy in damages for that breach and, other things being equal, those damages will be 'unliquidated' in the sense that any loss (however great) that the Employer can prove results naturally from the Contractor's breach will be recoverable. However, tenderers faced with that possibility might be deterred from tendering at all or, if they did tender, the price quoted might well be heavily inflated to cover the tenderer's perception of the risk. For this reason, the Contract prescribes 'liquidated damages' which will accrue automatically in the event that the Contractor fails to complete on time, and whether or not the Employer has suffered an actual loss either greater or less than the sum stipulated in the Contract — or, indeed, has suffered any loss at all.

Given a provision for liquidated damages, any failure on the Employer's part to terminate the Contract immediately the Contractor was in breach could be construed as condoning the breach and would almost certainly put time 'at large'. It would also destroy the Employer's right to liquidated damages and arguably also might prevent his recovering unliquidated damages (the legal authorities are less than clear on this point). However, the effect of the extension of time provisions is to preserve the right of the Employer to liquidated damages should the Contractor still fail to complete within the time so extended.

Liquidated damages must be a genuine pre-estimate of the actual loss likely to be suffered by the Employer should the Contractor be in breach. Thus a sum which is clearly excessive may be held to be a penalty and not a genuine pre-estimate, and penalties cannot be recovered under English law. On the other hand, commercial common sense will often induce the setting of a sum which is far too low, since otherwise no reasonable tenderer would bother to submit a tender. The setting of liquidated damages is thus very much a balancing act.

Finally, with regard to the granting of extensions of time, the Contractor is not only obliged but is entitled to complete the Works within the time prescribed. In other words, he is entitled to take up all the time available (provided that he finishes on time) whether or not he needs it or intends to use all of it. It must follow that any act of hindrance or prevention by the Employer will entitle the Contractor to an appropriate extension of time to put him back into the position he would have enjoyed had there been no such hindrance or prevention, whether he needs such an extension or not. In particular, any 'float' in the Contractor's programme is for his benefit and not the Employer's, and the latter has no right to appropriate it merely to mitigate the effect of his own misdeeds. On the other hand, there are a number of other causes of delay (such as those arising from unforeseen conditions or obstructions) which may be part of the risk borne by the Employer but which , while not the Contractor's fault, will only entitle the Contractor to such extension of time which in all the circumstances he can be shown to need.

Chapter 8

DEFECTS

Introduction

Most construction contracts contain a period after the completion of the Works during which the Contractor is responsible for putting right any defects that may appear. Such items as shrinkage cracks in plaster, joints in timber opening due to drying out, and the settling down of doors are imperfections that commonly appear within the first six months of use. The Contractor has been expected to return and put right these imperfections. The minor works contract has such a period, and it is referred to as the Defects Correction Period.

Defects Correction Period

> 5.1. 'Defects Correction Period' means the period stated in the Appendix which period shall run from the date certified as practical completion of the whole of the Works or the last period thereof.

The Certificate of Practical Completion is issued by the Engineer when the Works reach a state when they are taken or are fit to be taken into use or possession by the Employer (see Chapter 7). The certificate will specify the date upon which the Works reached that state (which will be earlier than the date when the certificate is actually issued), and it is from the date specified in the certificate that the Defects Correction Period runs.

It should be noted that where the Appendix to the Conditions of Contract requires parts of the Works to be handed over earlier than the date for completion for the whole of the Works, then it is the date of completion of the last of such parts that signals the start of the Defects Correction Period. The wording of the clause is a little unfortunate, but the intention is clearly that the Defects Correction Period starts when all of the Works have reached practical completion, and that there are not separate Defects Correction Periods for each part.

The length of the Defects Correction Period is specified in Item 10 of the Appendix to the Conditions of Contract, and Note 13(4) in the Notes for guidance advises on its duration.

> NG13(4) *Clause 5.1 (Defects Correction Period)*. This should normally be six months and in no case should exceed 12 months.

Under the Joint Contracts Tribunal (JCT) Building Forms of Contract, the maintenance period (which is the equivalent to the Defects Correction Period) is generally six months whereas under the ICE 5 Conditions of

Contract the period is normally 12 months. The minor works form permits either to be used but, as indicated in Note 13(4), the shorter period should normally be adequate.

Cost of remedying defects

> 5.2. If any defects appear in the Works during the Defects Correction Period which are due to the use of materials or workmanship not in accordance with the Contract the Engineer shall give written notice thereof and the Contractor shall make good the same at his own cost.

This requirement is limited to defects which are due to the use of materials or workmanship not in accordance with the Contract and does not, therefore, extend to defects arising from other causes such as design errors. Shrinkages and minor defects are covered by Clause 4.7, which excludes the repair of fair wear and tear resulting from the use of the Works by the Employer.

The Contractor is under no obligation to look for defects (but see Clause 2.3(b) for the Engineer's power to instruct him to do so) after the issue of the Certificate of Practical Completion, and his obligation to make good only starts when he is given written notice of any defects by the Engineer.

Remedy for Contractor's failure to correct

> 5.3. If any such defects are not corrected within a reasonable time by the Contractor the Employer may after giving 14 days' written notice to the Contractor employ others to correct the same and the cost thereof shall be payable by the Contractor to the Employer.

Clauses 4.7 and 5.2, taken together, impose a general duty on the Contractor to remedy defects whether these appear before Practical Completion (Clause 4.7) or during the Defects Correction Period (Clause 5.2). In either event, the Contractor has both a duty and a right to repair defects, provided that he does so within a reasonable time. The Employer has no right, either with his own staff or by another contractor, to remedy defects during the Defects Correction Period as, until that period has expired, the Contractor has an implied discretion to manage the work to be done, and may legitimately postpone repairs of early-appearing defects until he is sure that other similar defects are not about to appear.

The only exceptions to this will be if the Engineer is able to order immediate rectification under some other provision of the Contract (e.g. where the safety of the Works is involved, or if the Employer gives 14 days' written notice under Clause 5.3). However, the Clause 5.3 provision cannot be used until the Contractor has been given a reasonable time to do the work himself. What is a reasonable time will, of course, depend on the circumstances of each case. If the Contractor does the work himself, or if the Works are executed by others under permissible circumstances, then they will normally be carried out at the Contractor's expense.

Difficulty arises after the Employer has taken possession of part of the Works, since it may be arguable that the defects are due to fair wear and tear. The best practice in such cases is for the Contractor to carry out the repair, even when there is a doubt, and then claim reimbursement from the Employer. It will be necessary for the Contractor to show that the work involved was neither an unfinished item nor caused by materials or workmanship not in accordance with the Contract.

Completion certificate

5.4. Upon the expiry of the Defects Correction Period and when any outstanding work notified to the Contractor under Clause 5.2 has been made good the Engineer shall upon the written request of the Contractor certify the date on which the Contractor completed his obligations under the Contract to the Engineer's satisfaction.

The effect of the Engineer's certificate under Clause 5.4 is to terminate the general duty of the Contractor to repair defects and of the Employer's right to require the Contractor to do so. It also secures the release of the remainder of the retention money under Clause 7.5. It is therefore in the interests of the Contractor to obtain that certificate as early as possible. It is mandatory for the Engineer to issue such certificate upon request of the Contractor, provided that the Works have been completed to his satisfaction.

Unfulfilled obligations

5.5. Nothing in this Clause shall affect the rights of either party in respect of defects appearing after the Defects Correction Period.

Defects appearing after the expiry of the Defects Correction Period are not covered by the foregoing provisions, and Clause 5.5 makes it clear that the rights of the parties in respect of such defects remain unaffected. Although the Contractor cannot be compelled to remedy such defects, it may often be the most economical solution to try to persuade him to do so.

The Contractor remains liable for breaches of the Contract in respect of both materials and workmanship for the periods laid down in the Limitations Act 1980. In general, this means that actions for such breaches of the Contract must commence within six years after the work was carried out in a simple contract, and 12 years for a contract under seal (see legal comment at the end of this chapter).

Comparison with ICE 5

Maintenance Period under ICE 5 becomes Defects Correction Period under the minor works form. This purely semantic change creates a new term which accurately describes the matters with which it is concerned, namely the correction of defects after partial completion.

Under ICE 5 Clause 49(1), there can be different Maintenance Periods for different sections of the work, if the Engineer certifies that substantial completion took place other than on a single date. This is not so under the minor works form, as Clause 5.1 stipulates that the Defects Correction Period (singular) shall run from the date certified as practical completion of the whole of the Works, or the last period thereof. This latter arrangement should not substantially affect the Contractor for projects of limited duration.

Under the minor works form, the Contractor is given a reasonable time before the Employer is entitled to undertake the Works himself. Under ICE 5, Clause 49(4), the Contractor must 'fail to do any such work' before the Employer can undertake it. What is a reasonable time would be dependent upon the facts, but should be easier to ascertain than failure. Neither the Maintenance Certificate issued under Clause 61(2) of ICE 5 nor the Engineer's certificate at the end of the Defects Correction Period issued under Clause 5.4 of the minor works contract affects the rights of either party in respect of their other obligations.

Clause 49(5) of ICE 5, which covers the maintenance of temporary reinstatements of highways is not found in the minor works form, and any necessary requirement for the maintenance of temporary reinstatement of highways until such time as the local highway authority undertakes permanent reinstatement must be defined in the Specification.

ICE 5 also contains in Clause 50 authority for the Engineer to give instructions to the Contractor to search for defects, faults or imperfections. There is no similar clause in the minor works form, but similar powers are probably conferred upon the Engineer by Clause 2.3(b) (power of the Engineer to instruct any test or investigation).

Thus, with the exception of the separate maintenance periods for sections of the Works under the ICE 5, both contracts impose roughly similar liabilities upon the parties in respect of the correction of defects.

Comparison with ICE 6

Except that the Maintenance Period has now been renamed as the Defects Correction Period, thus bringing ICE 6 in line with the minor works form, there is no significant difference between ICE 6 and ICE 5 on these topics.

Legal comment

Two situations are, perhaps, worthy of further comment. The first arises during the Defects Correction Period if the Employer insists upon rectifying defects for which the Contractor is liable using his own resources or employing another contractor for that purpose while the Contractor remains able and willing to rectify the defect himself. In due course the Employer will seek reimbursement of the costs of rectification from the Contractor. However, such reimbursement should be limited to what it would have cost the Contractor to do the work himself. Thus actual costs, if excessive, must be reduced to a level which is reasonable in all the circumstances, allowances for overheads must be carefully scrutinized, and all profit elements must be rigorously excluded. In the end, the Engineer must decide, but that decision will inevitably be difficult, particularly as the parties will almost certainly be at loggerheads.

The second situation is that relating to latent defects, which are those which do not appear until after the Defects Correction Period has expired. As has already been stated, liability on contract for such defects will persist for six or twelve years (for simple contracts or contracts under seal respectively) from the date of the breach of contract giving rise to the defect in question. However, liability for defects can also arise in tort, and that liability will persist for six years after the date when the damage flowing from the defect was first suffered — in practice, from the date the defect first appeared.

But it is quite possible for a latent defect to appear, but not be noticed for some considerable time. In such cases the Latent Damage Act 1986 provides (in effect) that the right to sue for damages runs for six years from the date on which the defect first appears or for a further three years from the date upon which the defect was first discovered (or with due diligence ought to have been discovered). In addition, as many years might pass before the defect first appears, the Act prescribes a 'long-stop' period of fifteen years from the breach of duty giving rise to the defect, after which all actions at law in respect of it will be statute-barred. For all practical purposes, the tortious breach of duty in civil engineering projects will probably be the concurrent breach of contract.

Unfortunately, that is not the end of the story, since the Contractor will, at least in principle, be at risk from litigation (or arbitration) for even longer. The long-stop period really affects only the issuing of a writ (or the notice to refer to arbitration), after which the plaintiff has up to a year to serve it on the defendant. And even after service it can be many years before the issues are brought to trial. In the present state of the court lists, this means that judgment could be postponed until 20 or perhaps even 25 years after the original breach of duty. Moreover, should third parties be involved (for instance, suppliers or sub-contractors) from whom the defendant seeks contribution, such actions for contribution are not caught by the long-stop period.

Finally, the Latent Damage Act itself is defective in that it applies only to actions in negligence. Thus it remains open to a plaintiff to circumvent the long-stop provision by pleading his case in contract or in breach of statutory duty. Again, the fifteen years of the long-stop period is arguably far too long, particularly in the light of the ten year period used in other, analogous English legislation and for construction projects in other countries such as France (although in Scotland there is not only a 'short prescriptive period' of ten years but also a 'long prescriptive period' of twenty years).

For the construction industry, there is a potentially troublesome provision which overrides the long-stop period where there is 'fraud, misrepresentation or deliberate concealment'. This phrase was borrowed from the law relating to physical injury, and there can be no objection to the provision where fraud or misrepresentation is involved. However, 'deliberate concealment' must of necessity occur every time a formation is covered by a foundation, reinforcement is enrobed in concrete or masonry is plastered (the examples are, of course, endless), and such concealment is both innocent and an integral part of what the Contractor is required to do by the Contract. Unfortunately, every effort to persuade Parliament to retain the long-stop for *innocent* concealment was rejected by those having charge of the Bill. It is to be hoped that there will on some future occasion be some legislative opportunity to remove these blemishes from the statute.

Chapter 9

ADDITIONAL PAYMENTS

Introduction

The Contractor is entitled to receive additional payment for extra costs which he incurs as a result of matters for which the Employer carries the responsibility.

> 6.1. If the Contractor carries out additional works or incurs additional cost including any cost arising from delay or disruption to the progress of the Works as a result of any of the matters referred to in paragraphs (a) (b) (d) (e) or (f) of Clause 4.4 the Engineer shall certify and the Employer shall pay to the Contractor such additional sum as the Engineer after consultation with the Contractor considers fair and reasonable. Likewise the Engineer shall determine a fair and reasonable deduction to be made in respect of any omission of work.

The Contractor is thus entitled to additional payment for

- (i) the costs of additional works carried out, provided that they are not the result of some default by the Contractor (additional works will usually, but need not be, variations);
- (ii) additional costs incurred in constructing works which are not changed but their mode of construction is varied by an instruction of the Engineer; and
- (iii) the costs arising from delay and disruption from events for which the Employer is responsible.

Although the clause does not say so it must be implied that for the Contractor to obtain payment the costs must result from matters for which the Employer must bear responsibility under the Contract. These will include, but are not limited to, instructions and directions issued by the Engineer.

In all cases the Engineer is required to certify what is 'fair and reasonable'. It has to be remembered that these words take into account the interests of both the Employer and the Contractor. The Contract contains only two guidelines as to what is fair and reasonable and then only with regard to two specific causes of additional cost. These are to be found in Clause 6.2 for rates for varied work, and in Clause 3.8 for valuing the effects of finding unforeseen physical conditions or artificial obstructions.

> 6.2. In determining a fair and reasonable sum . . . for additional work the Engineer shall have regard to the prices contained in the Contract.

> 3.8 . . . then the Engineer shall certify and the Employer shall pay a fair and reasonable sum to cover the cost of performing any additional work or using any additional plant or equipment together with a reasonable percentage addition in respect of profit as a result of:-

(a) complying with any instructions which the Engineer may issue
and/or
(b) taking proper and reasonable measures to overcome or deal with the obstruction or condition in the absence of instructions from the Engineer

together with such sum as shall be agreed as the additional costs to the Contractor for the delay or disruption arising therefrom. Failing agreement to such sums the Engineer shall determine the fair and reasonable sum to be paid.

Otherwise the Engineer is on his own in determining what is fair and reasonable although, as will be seen later, some guidance can be obtained from legal precedent.

The words 'after consultation with the Contractor' imply that the Contractor has agreed to such consultation process and is therefore obliged to supply information on his rates and prices under consideration as to enable the consultation to be effective. This does not entitle the Engineer to a full breakdown of the Contractor's tender but merely to details of the individual rates under discussion to assist in establishing what is fair and reasonable by reference to the breakdown of direct costs, site overheads, head office overheads and profit reflected in the Contractor's tender rates.

Sources of additional payment

Variations

The Engineer is given wide power to vary the Works.

2.3. The Engineer shall have power to give instruction for:
(a) any variation to the Works including any addition thereto or omission therefrom:

The only restriction placed upon the Engineer is the implied condition that the variation must be within the scope of the Contract for which the Contractor tendered. This is defined in the brief description of the Works, or in the absence of such a description elsewhere in the contract documents the scope of the Works will be implied from the content of the Contract.

For instance, if the Contract were to be for the laying of a main sewer between two locations identified in the Contract, the Engineer could by variation instruct the Contractor to alter the levels of the invert, the diameter, number of manholes and even, perhaps, the precise route of the sewer, but he could not vary the Works to extend the sewer to serve a third location, since the Contract was simply to connect the first two.

Other instructions

Since the definition of the Works includes 'all work necessary for the completion of the Contract' the Engineer is able to give instructions with regard to the temporary works and methods of construction. Any of these may be included in the definition of 'additional work'.

There is neither a specific requirement for such instructions to be in writing, nor is there any procedure for the Contractor to confirm oral instructions. Clearly, good practice dictates that such instructions should be given in writing in order that there is proof that the order was given when it comes to agreeing payment. The reason for the absence of specific procedures is that on minor works contracts it is envisaged that neither party may have full time representatives on the Site and it is not practical to lay down a strict procedure for notifications. It is essential that an understanding be reached between the Contractor and the Engineer for each contract for the issue and acceptance of variation instructions if arguments about entitlement to payment are to be avoided later.

The absence of a provision for the Contractor to confirm oral instructions does not mean that the Contractor is not entitled to do so. If the Contractor writes confirming an alleged oral instruction that letter will constitute proof that the instruction was given unless repudiated within a reasonable time by the Engineer.

Assessing additional costs

Costs of additional works carried out

1. General

As the Contractor is, in effect, being asked to carry out more work than was envisaged when he submitted his tender, he is in principle entitled not only to the cost of the additional work but also a reasonable profit thereon. After all, if the extent of the additional work now ordered had been included when tenders were invited, tenderers would undoubtedly have included a profit element in pricing the work. Of course, in assessing what that profit element ought to be, the Engineer will need to find out, or at least make an intelligent estimate of, the profit margin actually included in the Contractor's tender rates and prices, bearing in mind that some items may well have been 'loaded' with profit while others in effect contained none.

2. Valuation on a rate basis

There are no rules covering the valuation of variations such as are found in most construction contracts. The Engineer has to value them on the basis of what is fair and reasonable. However limited guidance is given in Clause 6.2.

> 6.2. In determining a fair and reasonable sum under Clause 6.1 for additional work the Engineer shall have regard to the prices contained in the Contract.

The starting basis for valuing additional work is the assumption that the rates in the Contract are fair for similar work unless there are changed conditions involved which affect the cost of carrying out the work. This option will normally only be available when the Contract has been let on the basis of Measure and Value with Bills of Quantities or Valuation on a Schedule of Rates (Items 2(b) and 2(c) in the Appendix to the Conditions of Contract).

Unfortunately most methods of measurement, upon which the Bills of Quantity are based, refer to 'rates and prices'. Presumably, a rate is applied to a quantified item and a price is lump sum item. Clearly, the intention of the Contract is that the word 'prices' includes both.

The cost to the Contractor of executing work will vary considerably depending not only upon the specification of the work required, but also on the conditions under which it is carried out. On minor works contracts where there may be only one or two operations taking place concurrently the cost effects of variations may be considerable.

For instance, a variation to one part of the Works might affect one or more other parts not themselves varied. Thus haul-road distances could be affected for muck-shifting work to the Contractor's detriment, or the work which is varied might need larger or different plant and equipment to carry it out which, on a small project, might make the reversion to smaller plant for other parts of the project uneconomical. Temporary works might also be affected, particularly where repeated uses of, say, shuttering had been envisaged. The result may well be that the rates and prices in the Bill of Quantities are no longer appropriate for the Works as they are now to be carried out, and such inappropriateness can extend to work other than that varied.

Again, an urgent variation might involve stopping, or slowing down, some or all other activities, with labour, plant and equipment necessarily standing idle, or having their output reduced. It might also involve finance charges on materials awaiting use, and outgoings whilst the rates for the varied work are agreed. The costs of such stoppages, or less efficient working, being directly attributable to the variation, would be part of the overall cost of the variation and would clearly not be covered by the original rates for the work affected. However, such consequential costs are probably more conveniently dealt with as 'delay and disruption' (see below).

The bill prices may be used in one of three ways in determining what is fair and reasonable. These are as follows:

(i) The rate itself can be multiplied by a factor reflecting the degree of the change in the conditions under which the varied work was carried out in comparison with the conditions for the tendered work.
(ii) Where one or more individual components comprising the rate are varied, then only these components are varied to arrive at the new rate.
(iii) The levels of profit and overheads contained in the rate can be applied to the actual nett cost of carrying out the varied work to derive the new rate.

If, however, none of these methods is appropriate, the Engineer will be thrown back on his own resources and experience in arriving at what is fair and reasonable.

When fixing rates under Clause 6.1 the Engineer is acting as a valuer. In *Sudbrooke Trading Estate* v *Eggleton* (1982) Lord Diplock said 'a fair and reasonable price would be assessed by applying objective standards used by valuers in the exercise of their professional tasks'. The most common valuer of civil engineering works is the Contractor's estimator. It would seem reasonable, therefore, to conclude that the normal methods used in estimating will be appropriate for the valuation of variations. These will include (but are not limited to)

(i) deducing outputs from first principles,
(ii) referring to estimating price books (Spon's Civil Engineering Price Book, Wessex Data Base),
(iii) referring to estimating output books (Spence Geddes, Price Davies).

3. Valuation on a cost-reimbursement or daywork basis

If the Engineer requires that any variations or additional work ordered by him should be executed on a cost-reimbursement basis, that requirement must be included in the instruction, so as to enable the Contractor to prepare the necessary records substantiating his costs.

When the work is ordered to be executed on dayworks, it will be valued either at the rates contained in the Contract (if any) or in accordance with the 'Schedules of Dayworks carried out incidental to Contract Work' published by the FCEC (see Chapter 16).

> 2.5. The Engineer may order in writing that any work shall be executed on a daywork basis. Subject to the production of proper records the Contractor shall then be entitled to be paid in accordance with a Daywork Schedule included in the Contract or otherwise in accordance with the 'Schedules of Dayworks carried out incidental to Contract Work' issued by the Federation of Civil Engineering Contractors and current at the date the work is carried out.

Two points should be noted. Firstly, the Schedules are for works carried out incidental to the Contract and are only applicable to such works. If the daywork items become a substantial part of the Contract Works then the basis of the Schedules falls and the Contractor will be entitled to claim for a reassessment of the basis of his rates. Since daywork covers the direct cost of the work on a reimbursement basis, all that needs to be considered is the shortfall (if any) in the Contractor's recovery of overheads during the period when the work paid for on a daywork basis predominates.

The FCEC schedule rates contain an element in respect of overheads (approximately $12\frac{1}{2}\%$) which may be insufficient to cover the cost of the Contractor's site administration. To substantiate a claim for a higher percentage the Contractor will have to demonstrate to the satisfaction of the Engineer, that there was a shortfall between the the site overhead expenditure that he incurred while the daywork was being carried out and the recovery for overheads received through the daywork rates.

4. Valuing additional work in cost-plus-fee contracts

The direct costs of the work will be paid for in the same manner as the substantive contract work, but it may be necessary to reassess the fee if the additional work substantially increases the contract value. The fee usually covers the Contractor's profit, and under Clause 6.1 he is entitled to additional profit for additional work. The problem will not arise until the work executed exceeds that allowed for in the Contract. Provisional sums for contingencies will not be taken into account when assessing the fee, and should not be included when deciding whether the Contractor is entitled to an uplift in the fee.

5. Additional costs incurred by the Contractor

This source of additional payment in Clause 6.1 will cover additional costs resulting from Engineer's instructions which do not involve additional work. The Contractor may be involved (for example) in paying for services for other contractors employed by the Employer, for equipment and services for the

Engineer, for tests or other investigations or for providing additional insurances requested by the Employer.

The words are very wide and will cover any additional costs which the Contractor incurs providing that they do not arise as a result of a cause for which the Contractor is responsible.

The word 'cost' is defined in Clause 1.3.

> 1.3. 'Cost' (except for 'cost plus fee' contracts see Appendix) includes overhead costs whether on or off the Site of the Works but not profit.

In general, and unlike the position with respect to additional work carried out (see above), the Contractor will not normally be entitled to profit in addition to cost where the Works themselves are not changed. This is because the profit for which the Contractor bargained is already contained in the 'base' prices and rates for such work and, by recovering his additional costs (without profit thereon), he is being put back into the same commercial situation as he would have enjoyed had the additional costs not been incurred. It is sometimes argued that the Contractor should be entitled to a 'fair return' on all resources committed to the Works, but that is not — and never has been — the philosophy behind the ICE contracts.

Overheads, however, need to be treated differently. Site overheads will usually be time-based, so the effects of additional costs on site overheads will generally be linked to delay (see below). But head office overheads are calculated differently, usually on the basis of turnover or receipts, and therefore need to be considered separately. It is notoriously difficult to assess accurately the head office costs of a contractor incurred as a consequence of any particular event. A number of ways of assessing head office costs have been put forward by the authors of textbooks but none has received universal approval.

The difficulty arises because it is almost impossible to ascertain the costs of the services provided by a head office to any one contract. It is even more so for any one event. So various methods have been devised for allocating them as a contribution levied on a contract by proportioning the overall cost of running the head office to the annual turnover of the company and applying that proportion to the value of each contract. This will give a sum of money required from a contract as a contribution to running the head office. This sum can then be recovered from the Site either by fixed monthly amounts, or as a percentage of receipts. The former is the basis of the so-called Hudson formula, but it is the latter which is most appropriate for small contracts.

Since head office costs are usually included in the tender price as a percentage of cost, and additional payments contribute to turnover, it is logical to allow for head office costs by simply adding the percentage allowance made by the Contractor in his tender for his head office costs to the value of any additional payments made on the basis of costs incurred. Valuations made on the basis of tender rates will of course already contain an element for head office costs.

6. *Costs resulting from delay and disruption*

Variations ordered by the Engineer (Clauses 2.3(a) and 4.4(a)). The timing of the issue of variation orders can have a direct effect upon the costs of the work to the Contractor. If a variation order is issued when the affected work is in progress and causes the work to stop, cost will inevitably be incurred while the operatives and plant are redeployed on alternative operations. There may also be delay if the variation requires the purchase of materials which are on long delivery.

Even where a variation is ordered well in advance of the start of the work to be varied there may still be delay and disruption in that it may no longer be possible to carry out the whole or some significant part of the Works as originally programmed or in the originally intended sequence. 'Standing time' and 'slippage' may well be inevitable, whether on the work to be varied or in connection with other work, and to that extent the additional costs thereby occasioned will be recoverable from the Employer.

In either event, there will be a range of possible cost-inducing circumstances for which the Contractor will not be fairly reimbursed if he is paid only for work done at the rates in the Bill of Quantities. However, in such circumstances what is fair and reasonable may often conveniently be assessed by valuing work done at bill rates and making a separate assessment of the additional cost incurred as a result of the delaying factors.

It is arguable that delay and disruption costs arising from a variation order should, like the varied work itself, attract an element of profit. However, the logic behind such a statement is by no means clear, since if the work which is actually delayed and disrupted is not itself changed as a consequence of the variation, it can be argued with equal justice that profit should not be recoverable thereon, for the reasons already set out above. But each case will turn on its own facts, and it will be for the Engineer to determine whether and to what extent a profit element should fairly and reasonably be allowed.

Suspensions of the Works ordered by the Engineer (Clauses 2.3(c) and 4.4(a)). If, whilst the work is being executed, it is suspended on the instruction of the Engineer, the Contractor will suffer disruption while he finds alternative work for the operatives and plant which would otherwise have been employed to carry out the suspended work. He may also experience delay to the completion of the Works as a whole and may incur additional overhead costs. Following the philosophy of the Contract such additional costs will be recoverable by the Contractor without the addition of profit.

However, the Contractor may also be ordered to undertake protection or other additional works to safeguard the suspended work during the period of suspension. In evaluating this additional work the Engineer 'shall have regard to the prices contained in the Contract' in accordance with Clause 6.2. If he wishes it to be valued on a daywork basis then he must so inform the Contractor before the work is carried out, ideally when issuing the suspension order, otherwise the work should be valued on a market price basis.

Changes in the intended sequence of working ordered by the Engineer (Clauses 2.3(d) and 4.4(a)). Again the timing of the instruction will be all important to considerations of cost. The Contractor has an obligation to take all reasonable steps to avoid or minimize delay under Clause 4.4. A change in sequence notified in sufficient time to enable the Contractor to re-plan his operations will only result in additional payments if the Contractor finds it necessary to change the resources required for the work from those that he had intended to use, and he actually incurs additional cost in so doing. Additional costs resulting from changed resources will not normally attract profit, since the Works themselves have not been changed.

Costs of tests which fail to show work or materials not in accordance with the Contract (Clauses 2.3(b) and 4.4(b)). The Contractor will be paid for both the tests themselves and any consequential cost resulting from the taking of the tests, but not profit thereon. Taking tests and awaiting the results of the tests can seriously delay the site operations. For instance, if the Engineer requires bearing tests to be undertaken on the soil found at the bottom of an excavation all subsequent operations — blinding, reinforcement, formwork and base concrete — will be delayed. Similarly, the Engineer may suspect that concrete in a base pour is not in accordance with the Specification and may order both that cores be cut from it and that no further concreting work be carried out until the results of the tests have been obtained. If those tests show that the concrete accords with the Contract then the Contractor will be paid the costs of taking the cores and of any delay and disruption incurred.

Delay in receipt by the Contractor of necessary instructions, drawings or other information (Clause 4.4(d)). The Engineer is responsible for providing all necessary information to the Contractor under Clause 3.6. There is no requirement for the Contractor to request additional information necessary to enable him to complete the work and which is not in his possession. It is the Engineer's job to ensure that the Contractor has all the information he requires when he needs it because it is a principle of the Contract that all information will be supplied at the start of work. Any additional delay or disruption costs incurred by the Contractor as a result of shortage of information will be recoverable from the Employer but, again, without profit.

Failure by the Employer to give adequate access to the Works or possession of land required to perform the Works (Clause 4.4(e)). The provision of an access road suitable for the Contractor's plant equipment and operatives is the responsibility of the Employer. Any costs incurred by the Contractor because of the inadequacy of the access provided by the Employer will be recoverable, but without profit. These costs may include upgrading the access to enable it to carry heavy loads, negotiating additional wayleaves because the access is not wide enough, procuring right of air-space for cranes, and any licence payments involved. They may also include any delay to the Works resulting from the necessity for any of them.

Delay in the receipt of materials by the Contractor which are provided by the Employer under the Contract (Clause 4.4(f)). When the Employer supplies materials under the Contract the Contractor is only entitled to receive them at such times and in such quantities as are programmed in the Contract. If the Contractor gets ahead of programme and is delayed through lack of materials that the Employer is supplying then he is not entitled to the delay costs that he incurs thereby. However, if the Employer does not supply to programme all the resulting delay and disruption costs will be recoverable.

If the Contract has no specified delivery dates for such materials then the Contractor is entitled to assume that he will be supplied with them as he needs them, and if he is delayed or disrupted at all through a shortage of any Employer supplied materials he will be entitled to recover any additional costs incurred. As before, profit on such costs is not recoverable.

The consequences of artificial obstruction or physical condition (Clauses 2.3(e), 3.8 and 4.4(c)). In addition to certifying the cost of any delay or disruption to the Works, the Engineer is to certify, under Clause 3.8

> . . . a fair and reasonable sum to cover the cost of performing any additional work or using any additional plant or equipment together with a reasonable percentage addition in respect of profit as a result of:-
>
> (a) complying with any instructions which the Engineer may issue,
>
> and/or
>
> (b) taking proper and reasonable measures to overcome or deal with the obstruction or condition in the absence of instructions from the Engineer
>
> together with such sum as shall be agreed as the additional cost to the Contractor of the delay and disruption arising therefrom.

The work affected by the artificial obstruction or physical condition may or may not be changed as a result. If it is changed, and whether or not that change amounts to a variation of the work, the Contractor will be entitled to profit in addition to the extra cost of overcoming the obstruction or condition. If it is not changed, the normal rule is that additional cost only (i.e. without profit) is recoverable. However, Clause 3.8 expressly overrides the latter rule and allows profit even though the Works affected are not actually changed.

On the other hand, there is nothing in Clause 3.8 which expressly alters the position with regard to delay costs as defined above. It is arguable, therefore, that profit will not accrue to such costs.

Deductions for omitted work

> 6.1. . . . Likewise the Engineer shall determine a fair and reasonable deduction to be made in respect of any omission of work.

This proviso is in lieu of the clauses in many construction contracts which permit variation in the rates if significant changes of quantity occur, either as a result of variation orders, instructions or errors in the original Bill of Quantities.

When work is omitted from the Contract, it may have a greater effect upon the Contractor's operations and costs than simply not having to execute the work involved. There may be implications for the programme sequence of working, plant and labour on site, temporary works, mobilization costs already incurred, and general site overheads. For instance, if the Contractor has to open up a borrow pit, the cost of providing access and the removal of overburden is likely to be a fixed sum, regardless of the amount of material to be extracted. If the amount of borrow required is substantially reduced from that anticipated in the Contract, the Contractor, if paid at bill rates, only recovers a proportion of the costs allowed in his tender for opening up and preparing the borrow pit.

It therefore follows that if the Works are materially reduced by variation, the adjustment will not necessarily be a straight *pro-rata* reduction of the Contract charge for the work, although this will usually be the starting point for consultations with the Contractor. It will be for the Contractor to demonstrate to the Engineer that a straight pro-rata adjustment is not the fair and reasonable method of assessment of the value of the omitted work.

Lump sum contracts

Normally, there will be no breakdown of the lump sum suitable for valuing omitted work. The Engineer will have to agree with the Contractor a fair and reasonable amount. The starting point for such negotiations will be the direct costs of labour, material and plant not expended as a result of the omission.

Consideration of overheads and profit should be a separate consideration. Since overheads are largely time-based, there will have to be a time saving before a reduction should be made. With regard to profit, the philosophy of the Contract is that the work attracts profit, and by analogy, a reduction in work should lead to a reduction in profit. Profit should therefore be deducted at the rate that the Contractor allowed in his tender on any reduction made under the other headings, including site overheads.

However, the approach set out in the preceding paragraph needs to be reconsidered where the omission is of a very substantial proportion of the whole Contract Works. Thus, while site overheads are usually time-based, head office overheads are more likely to be based on turnover or receipts, and a substantial reduction in cash flow will, at least for a small Contractor, have consequences on the financing of overheads out of all proportion to the scale of the reduction itself. Again, it can be argued that, in looking to the Contract as originally tendered for as a source of profit (and thus of the return on working capital), the Contractor may have relinquished other opportunities to make profits elsewhere. Therefore a large reduction in the scope of the Contract Works should nevertheless fairly and reasonably entitle him to at least something approaching his previously expected level of profit. Indeed, if an omission is big enough in proportion to the original Works as let, it may be more appropriate for the parties to rescind the Contract and negotiate new contractual arrangements. In general, however, all that can firmly be said is that these are all matters of degree.

Measure and value contracts

Under these contracts, the Contractor is paid at the prices in the Contract for actual quantities executed. Omissions should therefore take care of themselves. However, the omission of substantial quantities of any item of work could lead to the Contractor failing to recover his costs adequately. The question then arises as to whether Clause 6.1 permits the Engineer to adjust the rates in relation to the omitted work. When the omission occurs as a result of an Engineer's instruction, it would seem clear that it does, but if the reduction in quantity results from an error in the original bill quantity, there may be some doubt.

Bills of Quantities, certainly under CESMM 2, do not themselves incorporate work into the Contract, they merely measure what is already in it — that which is detailed on the Drawings, described in the Specification, or included in the Conditions of Contract. An error in a Bill of Quantity is therefore arguably not an omission of work. However, this rather narrow view is clearly not what was intended by the clause.

Where errors in the Bill of Quantities do occur that substantially affect the recovery of costs by the Contractor, the Engineer should take this into account when assessing what is a fair and reasonable deduction.

Valuation based on a schedule of rates

Since there are no quantities included in a schedule of rates, the question of an omission of work will seldom arise. However, there may be contracts in which an indication of the quantities of work involved are given, and if there is a substantial reduction in any particular item, this should be considered an omission and the rate varied as may be fair and reasonable.

Valuation of a daywork or cost-plus-fees basis

The question of omission of an individual item of work will not arise. Only if there is a substantial reduction in the total value of the Contract Works will the Contractor have any claim, and this would be for breach of contract rather than the valuation of omitted work.

Comparison with ICE 5

ICE 5 is strictly a measure and value contract. This is clearly stated in Clause 56(1) and this is supported by the Form of Tender which contains no contract sum and Clause 55(1) which says the quantities are estimated and not to be taken as the actual and correct quantities of work to be executed by the Contractor. The minor works form, on the other hand, permits other forms of evaluation as defined in Item 2 of the Appendix to the Conditions of Contract. Comparison therefore can only be made on the basis of a measure and value contract and all the following comments should be read in this context.

Both contracts require that the measure and value be based on a priced Bill of Quantities which is drawn up in accordance with a standard method of measurement which is to be specified in the Appendix to the Form of Tender in ICE 5 and Appendix to the Conditions of Contract in the minor works form.

ICE 5 states in Clause 57 that the Bills of Quantities are deemed to have been prepared in accordance with the method of measurement and provides for any errors of omission to be rectified under Clause 55(2). The minor works form incorporates both these, but by implication only.

If the quantities differ substantially from those in the Bill of Quantities ICE 5, Clause 56(2) permits the Engineer to fix alternative rates where it is considered that the bill rates are unreasonable or inapplicable as a result of the changed quantity itself. The minor works form has no such express provision but neither has it any express disclaimer concerning the accuracy of the quantities. It is left to the Engineer to value the Works on a fair and reasonable basis under Clause 6.1. Payment to the Contractor for additional work executed is based on the principle of *quantum meruit* (literally 'what it is worth') and that implies a market price.

The valuation of variations under ICE 5 is carried out in the three-stage manner found in most construction contracts — where work is similar and executed under similar conditions to the work priced in the Bill of Quantities it is valued at bill rates; where the work is not of similar character or is not executed under similar conditions, then the rates in the bill are used as the basis of rates for valuation; and if that cannot be done then a fair valuation has to be made.

The minor works form reduces this three-stage process to one — the sole criteria being that the Engineer shall certify what he considers is fair and reasonable having regard to the prices contained in the Contract. In both contracts the Engineer values variations after consultation with the Contractor. In practice there should be little difference in the way that variations are valued under each contract.

Under ICE 5, if the Contractor and the Engineer cannot agree rates, then the Engineer fixes them and thereafter the Contractor has 28 days in which to notify the Engineer that he proposes to claim a higher rate than the one fixed. Under the minor works form agreement is not necessary and the Engineer fixes the rate in any event. The Contractor can then raise objection to any of the Engineer's values at any time thereafter until 28 days after the Engineer has issued the final certificate under Clause 7.7, after which he will be time-barred.

The payment for additional costs incurred as a result of Engineer's instructions and directions is very much simpler under the minor works form. The Contractor is entitled to both the direct costs and any costs resulting from delay and disruption on the basis of what is fair and reasonable under Clause 6.1. This is markedly different from the approach adopted in ICE 5 where the Contractor is entitled to additional costs resulting from Engineer's instructions generally under Clause 13.1, and throughout the Contract there are various specific clauses, each with slightly different payment conditions in respect of additional cost.

The additional cost incurred as a result of delay and disruption is based upon the legal principle of 'damages' under ICE 5, and is therefore valued without profit. This is so under

Clause 7(3)	Late issue of instructions,
Clause 12(3)	Effects of encountering unforeseen physical conditions or artificial obstructions,
Clause 13(3)	Effects of Engineer's instructions,
Clause 14(6)	Consent to Temporary Works and methods of construction,
Clause 27(6)	Effects of the Public Utility Street Works Act 1950.
Clause 31(2)	Services to other Contractors,
Clause 40	Suspension of the Works, and
Clause 42	Late possession of the Site.

However, the actual wording varies between clauses. For delay in providing necessary information the Contractor is paid 'such costs as may be reasonable' (Clause 7(3)) whereas if the Engineer's consent to the Contractor's proposed methods of construction is unreasonably delayed or he requires changes to them that the Contractor could not have foreseen then the Contractor is entitled to be paid for 'such costs incurred as the Engineer considers fair in all the circumstances' (Clause 14(6)). If it was street work that caused delay then the Contractor is entitled to additional costs as the Engineer 'considers to have been reasonably attributable to such delay' (Clause 27(6)). And if the Works are suspended then the Contractor is paid 'the extra cost incurred in giving effect to the Engineer's instructions' (Clause 40(1)). 'Such costs as may be reasonable' are awarded for late possession of the Site under Clause 42(1).

Under the minor works form, however, virtually all the provisions on additional payments have been concentrated into Clauses 6.1 and 6.2, thereby avoiding the ICE 5's plethora of slightly differing definitions. The result is that the principles animating these provisions are (or should be) sufficiently clear to avoid any but the most intransigent difficulties in assessing such payments.

Turning now to additional payments in respect of unforeseen conditions or obstructions, ICE 5, Clause 12(3) is broadly the same as minor works Clause 3.8.

In conclusion it may, perhaps, fairly be said that the valuation of additional payments in practice should be broadly the same, whichever form of contract is used. The main practical difference is that the vast amount of detail contained in ICE 5 is greatly simplified under the minor works form.

Comparison with ICE 6

ICE 6 has followed the minor works precedent of collecting all like provisions in one place instead of the scattering of similar provisions across the text as seen in ICE 5. The result of this has been a much greater uniformity of both language and treatment in ICE 6 and a sharpening of the distinction between payment for additional work (which in principle attracts profit) and payment for additional cost of doing the same work, or for delay and disruption, neither of which will normally attract profit. However, the slightly anomalous position with respect to unforeseen conditions or obstructions is introduced in that all costs reasonably incurred do attract profit.

In short, therefore, there is even less difference between the minor works form and ICE 6 than between the former and ICE 5.

Chapter 10

PAYMENT

Introduction

The Contract allows for various forms of payment to be used and is not confined to the normal measure and value basis with Bills of Quantities. However, whichever method is used, the Contractor is entitled to monthly progress payments for the value of works executed to date and for materials delivered to the Site but not yet incorporated into the Works.

Valuation of the Works

7.1. The Works shall be valued as provided for in the Contract.

Various methods of payment can be specified in Item 2 of the Appendix to the Conditions of Contract:

2. The payment to be made under Article 2 of the Agreement in accordance with Clause 7 will be ascertained on the following basis. (The alternatives not being used are to be deleted. Two or more bases for payment may be used on one Contract.)
   (a) Lump sum
   (b) Measure and value using a priced Bill of Quantities
   (c) Valuation based on a Schedule of Rates (with an indication in the Schedule of the approximate quantities of major items)
   (d) Valuation based on a Daywork Schedule
   (e) Cost plus fee (the cost is to be specifically defined in the Contract and will exclude off-site overheads and profit)

Each of these bases for payment stipulates a different method of evaluation of the amounts that the Contractor should be paid in both the monthly payments under Clause 7.2 and the final account under Clause 7.6.

Lump sum

A lump sum contract does not require a Bill of Quantities. The Contractor will usually have to take off the quantities from the Drawings and Specification, and will be responsible for their accuracy.

Lack of a breakdown of the tendered lump sum into amounts for the individual items of work may cause difficulty in calculating the value of the work carried out for the purpose of monthly payments. There are three ways in common usage for overcoming this problem.

1. Schedule of stage payments

The basis of a schedule of stage payments is that the Contractor will be paid stated instalments of his contract sum when he achieves certain identifiable stages in the construction of the Works. These stages can either be time-related, or based upon physical progress (or 'milestones').

If the basis is to be time-related the Contractor will be asked to supply with his tender a breakdown of his lump sum into the monthly instalments that he will require each month. He will then be entitled to payment of each monthly instalment providing that his progress is in accordance with an agreed programme. The Engineer needs to check prior to the award of the Contract that the instalments proposed by the Contractor are in line with his estimate of the value of the works proposed to be completed by the end of each month.

One advantage of this method to the Employer is that he knows reasonably accurately the amount of money that he will be required to find each month. The disadvantage for the Contractor is that if he gets ahead of programme he does not get paid sums additional to the proposed instalments.

Alternatively, the Contractor will be asked to supply a schedule of payments which he will require to be paid when he achieves completion of particular stages in the Works. These stages in the Works are often called 'milestones'. For instance, for a bridge the milestones might be excavation of the foundations, the concreting of the foundations, the construction of the piers, the erection of the support for the superstructure, the concreting of the deck, and finally the completion of the bridge. The Engineer would certify the appropriate payment when each milestone is reached.

2. Contractor's Bill of Quantities

The Contractor may be asked to supply what is known as a 'Contractor's Bill', which is a list of the major quantities of the work involved against which a price is inserted by the Contractor, the total of the bill amounting to the lump sum. Such a bill will not be measured in accordance with a standard method of measurement but will cover just the quantities of the bulk items such as excavation, concrete and finishings. The Contractor will need to include for all his operations within these bulk quantities, and his monthly statement will be based upon the amount of these bulk quantities that have been executed. If this method is proposed the Engineer should check both that the bulk quantities are approximately correct and that the rates against them are reasonably related to the value of the work covered by them.

3. Agreement between Contractor and Engineer

In the absence of either of these methods the Engineer is left to agree with the Contractor the value of the work executed each month. This can be done by agreeing a percentage of the Contract Works executed so far and certifying that percentage of the lump sum.

Variations may prove expensive, and agreement on their pricing difficult, because there will be no rates in the Contract as a basis of negotiation even if a Contractor's Bill is proposed. However, this difficulty should not be exaggerated since there are always the standard pricing books to fall back on.

Commentary

Measure and value using a Priced Bill of Quantities

The measure and value contract will have a Bill of Quantities which will have been drawn up by the Engineer in accordance with the rules of a standard method of measurement which is to be defined in Item 3 of the Appendix to the Conditions of Contract.

> 3. Where a Bill of Quantities or a Schedule of Rates is provided the method of measurement used is
>
> ....

The effect of this provision is that the Bills of Quantities are warranted by the Employer to have been compiled in accordance with the method of measurement, and any errors or deviations from the method (unless detailed in bills) will need to be corrected by the Engineer, and the Contractor will be paid on the basis of the corrected bills.

It has to be remembered that the method of measurement must be read as a whole, and there are provisions where the Contractor has measured and paid for items of work which he may not carry out (i.e. back shutters for concrete walls, top shutters, etc.). Rates for items of work which are omitted from the Bills in error will have to be appraised not on the basis of the cost of the work to the Contractor but what is a reasonable market rate for the type of work described.

Valuation based on a Schedule of Rates

This alternative will only have been adopted when the actual quantities of work cannot be measured with any degree of accuracy. Additional work is to be expected and most of it will be evaluated in accordance with the schedule. However, the Contractor can only price works on the basis of what is expected at the time of tender, and if the conditions under which additional work is carried out differ from that basis, the Contractor will be entitled to have the work valued not on the basis of the Schedule of Rates, but on what is fair and reasonable.

Valuation based on a Daywork Schedule

When the Contract is let on the basis of a Daywork Schedule the Contractor will be required to record the costs of labour, materials and plant used. He will also be required to supply the necessary supporting documents for verification by the Engineer. The Contractor will then be paid in accordance with the arrangements for dayworks as prescribed by the Contract, as discussed in Chapter 16.

Cost plus fixed fee

Where this method is adopted it is necessary to define clearly the basis for the assessment of 'cost' and what the Contractor is required to cover in his 'fee'. There are several published contracts which will assist in defining cost and the JCT (Joint Contracts Tribunal) Fixed Fee Form of Prime Cost Contract is a useful source of such a definition. The FCEC daywork schedule (see Chapter 16) is another.

The fee can be a fixed lump sum, a percentage of the eventual cost, or variable according to some form of agreed target. However, since the work involved is likely to be low in value, the first two alternatives should prove adequate for use with this form of contract.

For interim payments cost and fee must be considered separately. The payment for cost will follow the procedure required for dayworks — the Contractor will submit his time and wages sheets, material and plant invoices, and all receipts and other records of payments for site overhead costs for the evaluation of the site cost. The fee will be paid in the manner defined in the Contract, which may be either as equal monthly instalments over the period of the Contract, or proportionately to cost.

Monthly payments

Valuation and certification

Payment is by monthly certificate. The Contractor is required to submit a monthly statement showing the amount that he considers should be included in a monthly certificate.

> 7.2. The Contractor shall submit to the Engineer at intervals of not less than one month a statement showing the estimated value of the Works executed up to the end of that period together with a list of any goods or materials delivered to the Site and their value and any other items which the Contractor considers should be included in an interim certificate.

The requirement is only for the 'estimated value' of work to be included in the statement. This does not necessarily mean that a full measurement has to be carried out each month. Any method that will produce a sufficiently accurate value can be used, such as taking the amount of the total value of measured work that is proportional to the time that has elapsed of the total contract period.

The intention of the words 'any other items' is explained in Note 12 of the Notes for guidance

> NG12. The reference to 'any other item' in Clause 7.2 is to permit the Contractor to include in interim valuations other amounts to which he considers himself entitled such as goods or material vested in the Employer but not yet delivered to the Site or the value of temporary works or constructional plant on the Site for which there is separate provision for payment in the Contract

The amounts for additional payments, although not mentioned in Note 12, will be covered by the words 'estimated value of the Works' because they will include all amounts payable under the Contract.

Whilst there is no specific clause covering the vesting of materials or plant in the Employer, it is nevertheless reasonable for the Contractor to receive advance payment for items that are both specially manufactured for the Contract and whose cost is high in relation to the contract value. Before such an advance is made the Engineer must come to a satisfactory agreement with the Contractor to ensure that ownership of the the materials or plant has been vested in the Employer.

Commentary

The definition of 'Site' excludes land provided by the Contractor for the purposes of the Contract and therefore will preclude payment for materials (as 'materials on site') which may have been delivered to such land.

Payment

The Engineer is required to certify and the Employer to pay the Contractor the sum the Engineer considers properly due within 28 days of the delivery of the monthly statement by the Contractor.

7.3. Within 28 days of the delivery of such statement the Engineer shall certify and the Employer shall pay to the Contractor such sum as the Engineer considers properly due less retention at the rate of and up to the limit set out in the Appendix. Until practical completion of the whole of the Works the Engineer shall not be required to certify any payment less than the sum stated in the Appendix as the minimum amount of interim certificate. The Engineer may by any certificate delete correct or modify any sum previously certified by him.

The Engineer and the Employer must agree procedures which will ensure that the 28 days can be met, including the amount of time each requires to fulfil his duties. Nevertheless, should the 28-day limit be exceeded, there is provision for the payment of interest on late payments in Clause 7.8 which is at 2% above base lending rate at the bank specified by the Contractor under Item 14 of the Appendix to the Conditions of Contract (which will be completed by the Contractor when he submits his tender).

Minimum certificate

The minimum amount of an interim certificate is given in Item 13 of the Appendix to the Conditions of Contract. It is suggested in Note 13(6) of the Notes for guidance that this should be '10% of the estimated final Contract value rounded off upwards to the nearest £1000'. It is also important to note that this limit only applies until the issue of the certificate of practical completion for the whole of the Works, and thereafter the Engineer is required to certify and the Employer to pay any sums that may be due, however small.

Correction of certificates

Under Clause 7.3 the Engineer can correct or modify any sum previously certified by him. Interest would be payable on any sum under-certified in a previous certificate, provided that the reasons for the under-certifying were not the responsibility of the Contractor.

Retention money

The rate of retention is detailed in Item 11 of the Appendix to the Conditions of Contract and the limit is specified in Item 12. The Engineer must fill in these two items before tenders are invited. Guidance on the these two figures is given in Note 13(5).

NG13(5). Clause 7.3 (Rate of retention and limit of retention). The rate of retention should normally be 5%. A limit of retention has to be inserted in the Appendix and this should normally be between the limits of $2\frac{1}{2}$% and 5% of the estimated final Contract value.

It should be noted that the limit is to be expressed as a sum of money and not a percentage.

Payment of retention money

One-half of the retention is to be certified by the Engineer and paid to the Contractor within 14 days of the date on which the Engineer issues the Certificate of Practical Completion for the whole of the Works (See Clause 4.5), and the last half within 14 days of the Engineer issuing his completion certificate signifying that all defects have been corrected.

7.4. One half of the retention money shall be certified by the Engineer and paid to the Contractor within 14 days after the date on which the Engineer issues a Certificate of Practical Completion of the whole of the Works.

7.5. The remainder of the retention money shall be paid to the Contractor within 14 days after the issue of the Engineer's certificate under Clause 5.4.

It should be noted that the first half of the retention money is certified by the Engineer as expressly provided for (Clause 7.4), while the second half is not. Nevertheless, it would be both proper and desirable for the Engineer to certify this also, preferably at the same time as he issues his certificate under Clause 5.4.

Final account

The Contractor is required to prepare the final account within 28 days of the Engineer certifying that the Contractor has corrected all defects and completed his obligations under the Contract. The Engineer then has 42 days in which to verify the statement and issue the final certificate. The Employer must then pay within a further 14 days. This means that the Contractor is entitled to receive full payment within 84 days of completing all work required at the end of the Defects Correction Period.

7.6. Within 28 days after the issue of the Engineer's certificate under Clause 5.4 the Contractor shall submit a final account to the Engineer together with any documentation reasonably required to enable the Engineer to ascertain the final contract value. Within 42 days after the receipt of this information the Engineer shall issue the final certificate. The Employer shall pay to the Contractor the amount due thereon within 14 days of the issue of the final certificate.

The final certificate is conclusive evidence as to the sum due to the Contractor unless either party invokes the disputes procedure defined in Clause 11 within 28 days of the issue of the final certificate (see Clause 7.7). Although it does not say so, it is probably implied that 'issue' means 'issue to the Contractor' and the time-barring will start from the time that the final certificate is received by the Contractor.

Final certificate

The effect of issuing the final certificate is that the Contractor is then barred from submitting further invoices in respect of the Contract. If the Contractor does not agree with the final certificate then he has 28 days within which to issue a Notice of Dispute under Clause 11.2 and the matter is then referred to arbitration.

7.7. The final certificate shall save in the case of fraud or dishonesty relating to or affecting any matter dealt with in the certificate be conclusive evidence as to the sum due to the Contractor under or arising out of the Contract (subject only to Clause 7.9) unless either party has within 28 days after the issue of the final certificate given notice under Clause 11.2.

This is a time-bar clause, and both parties (and the Contractor in particular) must be careful to check the final certificate when it arrives. After the 28 days it will be too late to complain. There is no express provision in the Contract requiring notices or certificates to be sent either to the Contractor or the Employer, but it will, it is suggested, be good practice to send copies to both parties whenever any notice or certificate is issued. If this is done, then in the case of a dispute, the 28 days will only be deemed to start running from the day that it is, or should have been received by each of the parties.

. . . shall save in the case of fraud or dishonesty relating to or affecting any matter dealt with in the certificate . . .

Dishonesty or fraud will include the Contractor's knowing submission of false information for the purposes of verifying the final account, the giving and taking of bribes and the Engineer's knowing omission of items from the final account. The final certificate carries no implication with regard to the quality of the work that has been completed, and is not a certificate of acceptance that all or any of the work complies with the requirements of the drawings or the specification.

Interest on overdue payments

Interest is payable on overdue payments by the Employer.

7.8. In the event of failure by the Engineer or the Employer to make payment in accordance with the Contract the Employer shall pay to the Contractor interest on the amount which should have been certified or paid on a daily basis at a rate equivalent to 2% per annum above the base lending rate of the bank specified in the Appendix.

Interest on an overdue payment is not an amount to be certified by the Engineer. The Contractor should raise a separate invoice for interest and send it direct to the Employer. The rate payable is 2% above the base lending rate of the bank specified in the Appendix to the Conditions of Contract (Item 14), and will normally be the Contractor's bank. It is likely that 2% is the average amount above the base lending rate that the Contractor will have to pay on his overdraft, and thus more or less in line with normal commercial practice.

The Engineer and the Employer between them have 28 days in which to certify and pay. They must sort out between themselves how long each can be allowed to perform his function.

Interest should not be confused with financing charges which are an element of cost in a claim for additional payment. Interest is due if the Employer fails to pay by the due date, whereas financing costs are the cost of providing the money needed to pay for the additional work or delay and disruption, and will arise (if at all) whether or not the Employer is late in paying. The rate used for calculating financing charges will be the rate at which the Contractor borrows the money necessary to finance the additional work, whereas interest on late payments will be at the rate in the Contract (2% above the base lending rate of the bank named in the Appendix to the Conditions of Contract).

There are different legal requirements in relation to interest and financing charges in Scotland, and reference should be made to Chapter 15 for comments on these matters in Scotland and other jurisdictions.

Value-added tax

> 7.9. In addition to sums due otherwise to the Contractor under the Contract and notwithstanding any time for payment stipulated in the Contract the Employer shall pay to the Contractor any value-added tax properly chargeable by the Commissioners of Customs and Excise on the supply to the Employer of any goods and/or services by the Contractor under the Contract.

All rates quoted by the Contractor are exclusive of VAT, and to the extent that any work attracts VAT this will be added to the amounts otherwise due to the Contractor under the Contract. In producing budgets for the Employer the Engineer should take any liability for VAT into account and add it to the sums payable to the Contractor.

It is normal for the Contractor to assess at the time of his monthly statement the amount of VAT that he believes is payable by the Employer. VAT should not, however, be included in either the Contractor's monthly statement, as submitted to the Engineer, or in interim or final certificates certified by the Engineer. These should show the amount due for payment nett of tax.

On receipt of each certificate, the Contractor should then invoice the Employer direct for the full amount of the certificate plus VAT. Notwithstanding anything that the Contract may provide, Contractors remain liable to account to the Commissioners of Customs and Excise for sums due in VAT on work they perform or services they provided, but this is only of concern if the Employer defaults.

Comparison with ICE 5

The regime for payment in both contracts is broadly similar. The Contractor is required to submit a monthly statement to the Engineer and thereafter the Engineer and the Employer between them have 28 days in which to certify and pay. Both contracts enable the Contractor to receive payments on account of both work and material on site. Such differences as there are between the two contracts are of marginal importance. Payment is covered under Clause 60(1) of ICE 5 which requires the Contractor to submit a statement at the end of each month whereas Clause 7.2 of the minor works form requires statements at intervals of not less than one month.

The breakdown of the monthly statement required under Clause 60(1) is into four items:

(i) the measured work,
(ii) materials on Site,
(iii) materials manufactured but not yet delivered to the Site, and
(iv) all other matters to which the Contractor considers himself entitled.

The first two items are covered in Clause 7.2 of the minor works form and the last by the words 'and any other items which the Contractor considers should be included in an interim certificate'. These words give scope for the inclusion in a monthly statement of materials 'not on Site' and any other matters for which costs have been incurred. They do not necessarily imply that the Contractor is entitled to payment for them. If not expressly covered in the Contract, this is a matter for negotiation between the Engineer and the Contractor.

Under both contracts there is provision for a minimum interim certificate and in both contracts this requirement falls away upon the issue of the completion certificate.

The Engineer is entitled to reassess the value of work measured in any certificate. The requirements of ICE 5, Clause 60(7), are more elaborate than those found in Clause 7.3 of the minor works form because of the inclusion in the former of provisions for Nominated Sub-contractors.

Under both contracts there is provision for retention to be withheld from amounts otherwise due to the Contractor under interim certificates. These are covered by Clauses 60(4) and 60(5) in ICE 5 and Clauses 7.4 and 7.5 in the minor works form. The major difference between the two contracts is that the amount of the retention is specified in ICE 5 whereas it is left to the Engineer to stipulate it in the Appendix to the Conditions of Contract under the minor works form.

Under both contracts half the retention is released 14 days after the issue of the completion certificate and the other half 14 days after the issue of the maintenance certificate or its equivalent under the minor works form.

Interest is recoverable under both contracts by the Contractor at a rate of 2% above a specified base rate. In the case of ICE 5 the base rate is that at which the Bank of England will lend to a discount house having access to the discount office of the bank but under the minor works form it is a specific bank named in the Appendix to the Conditions of Contract. The intent is the same even if the mechanism is slightly different.

In both contracts the Contractor is required to submit a final account to the Engineer. Under Clause 60(3) of ICE 5, this has to be done within three months of the issue of the maintenance certificate whereas, under Clause 7.6 of the minor works form, the Contractor has only 28 days, which reflects the smaller task involved. The Engineer then has three months under ICE 5 or 42 days under the minor works form from receipt of the final account to issue the final certificate. Both of these periods will be dependent upon the Contractor submitting satisfactory documentation to enable the Engineer to verify the certificate.

The effect of the final certificate is the same under both contracts. It represents the sum that, in the Engineer's opinion, is finally due to the Contractor under the Contract. However, once issued, procedures for objection do differ. Under the minor works form the Contractor has 28 days in which to object by giving a Notice of Dispute whereas under ICE 5 the procedure for objection is first to refer the matter for reconsideration by the Engineer under Clause 66(1), who must give his decision thereon within three months. Thereafter the parties have a further three months to refer the matter to arbitration if they so wish. There is no time limit within which the Contractor has to refer the final certificate to the Engineer but after he has done so the strict timetable of Clause 66 comes into play.

The contrast between the two clauses for value-added tax is striking. Clause 70 of ICE 5 covers 75 lines of print and this is reduced to six in Clause 7.9. Nevertheless, the nett effect of both clauses is the same, namely that the Employer is liable to pay to the Contractor any value-added tax properly assessed by the Commissioners of Customs and Excise. The difference between the two probably lies in the fact that when the ICE 5 clause was first written nobody really understood the workings of value-added tax whereas by 1988 the problems had been ironed out and draughtsmen of the minor works form would be satisfied with simple wording.

Comparison with ICE 6

There is little practical difference between ICE 5 and ICE 6 so far as the payment clauses are concerned.

The definition of cost has been expanded and ICE 6 states that cost means 'all expenditure properly incurred or to be incurred whether on or off the Site including overhead finance and other charges properly allocatable thereto but does not include any allowance for profit'. This wider definition may provide guidance for the meaning of 'additional cost' in Clause 6.1 of the minor works form. It is also now expressly stated that where recovery is on the basis of cost in ICE 6 profit shall be added 'in respect of . . . any additional permanent or temporary work'. The disputes procedures do, however, differ from those in ICE 5.

Legal comment

In *Morgan-Grenfell* v *Seven Seas Dredging* (1990) it was held that interest pursuant to ICE 5 Clause 60(6) was recoverable from the date when, in the arbitrator's opinion, additional payments ought properly to have been certified. In addition, although the judgment was a little obscure on the point, it seems that such interest would also be compounded. This last point is at variance with the situation as usually understood, namely that interest awarded 'pursuant to statute', like interest on a judgment debt, is always simple and not compounded (although arbitrators have often adopted a rate of interest such that the total amount of interest awarded approximates to that under a compounded commercial rate).

This back-dating of interest is, of course, in line with commercial common sense, although *Morgan-Grenfell* v *Seven Seas Dredging* is the first case in which it has been given judicial support. As for compounding, it is unclear from the judgment how this was to be applied. However, the final version of ICE 6 has since been amended (Clause 60(7)) to make express provision that interest on failure to pay or to certify shall be compounded monthly.

No such express provision appears in the minor works form. Nevertheless, it is reasonable to suppose that the *Morgan-Grenfell* principle is of general application and, if so, it should be permissible to operate the minor works provisions on interest with monthly compounding.

Chapter 11

ASSIGNMENT AND SUB-LETTING

Introduction

Assignment and sub-contracting are often confused one with the other. The section on legal comment at the end of this chapter sets out to make the essential differences between the two terms clear. Both, however, involve contractual relationships with a third party, and for this reason only it seems convenient to consider them together.

It is unfortunate that the minor works form perpetuates the use of the term 'sub-letting' instead of the correct term 'sub-contracting'. Although sub-letting is indeed a form of sub-contracting, its use should properly be limited strictly to contracts for a legal interest in land. Thus, for example, the lease of a house entitles the lessee to the 'quiet possession and enjoyment' of the property demised, but there can be nothing of this nature to be gained from entering into a sub-contract for civil engineering services.

The misuse of the term sub-letting probably originated within the building industry where entrepreneurs frequently construct buildings and then lease them out to tenants. There would appear to be no such excuse in civil engineering, other than the fact that books of legal authority were at first mainly concerned with building and only later widened their scope to take in civil engineering. Be that as it may, it is to be hoped that in any future revision of the minor works form the opportunity will be taken to substitute the correct terminology.

Finally, and for the purpose of this chapter, contractual burdens are, for the Contractor, those obligations necessary in order to complete the Works and the other risks borne by him in connection therewith and, for the Employer, the obligation to pay the contract price, provide the Site and bear the Excepted Risks. Similarly, contractual benefits are the Contractor's right to be paid and the Employer's right to receive the completed Works on time and in good order.

Assignment

Under the minor works form neither party is entitled to assign either the whole or, indeed, any part of the Contract without the prior written consent of the other party.

> 8.1. Neither the Employer nor the Contractor shall assign the Contract or any part thereof or any benefit or interest therein or thereunder without the written consent of the other party.

Examples of situations where assignment might be requested are where the

Contractor wishes to assign the payments made against the monthly certificates to a factoring house, or where the Employer wishes to transfer the ownership of the project Works to an investment fund or some other purchaser.

In practice, requests for assignment should be rare. When they occur, they are likely to be resisted mainly because the objecting party does not wish to deal with the prospective assignee. Thus a Contractor might well have declined to tender at all if he had known that the assignee was to be the Employer, and similarly the Employer might have accepted the assignor Contractor's tender primarily on the basis of that Contractor's reputation and expertise, which the proposed assignee may or may not possess. However, assignments of the Contractor's rights to receive payment against certificates is unlikely to give rise to trouble, save to the extent that a request of this kind may be an early indication of Contractor insolvency.

Sub-letting or sub-contracting

8.2. The Contractor shall not sub-let the whole of the Works. The Contractor shall not sub-let any part of the Works without the consent of the Engineer which consent shall not be unreasonably withheld.

In contrast to assignment, sub-contracting is now widespread in the construction industry, many main contractors doing little work themselves beyond overseeing a host of sub-contractors who actually construct the Works. This situation is admittedly less likely to obtain on a minor works project, since the Works themselves will usually be small enough to make any significant sub-division of little advantage to the Contractor. Nevertheless, there will often be parts of the Works of a preliminary or specialist nature which the Contractor will find convenient to hive off to others.

While the Contractor must obtain the Engineer's consent (which should preferably be in writing) to any proposed sub-contracting, such consent is not unreasonably to be withheld. Adequate reasons for withholding consent would include well-founded doubts as to the proposed sub-contractor's competence or financial viability or where, before the Contract was awarded, it had been made clear to the Contractor that the Employer was relying on that Contractor's own experience, plant, labour or reputation. Again, consent could be withheld temporarily until such time as the Contractor's proposed means of supervising and controlling the sub-contractor's operations had been properly explained and agreed. Other than in these and similar instances, however, the Engineer should be slow to withhold consent, since Clause 8.2 does not give either the Engineer or (through him) the Employer a blanket veto to satisfy their own whims and fancies (but see below).

The whole Works not to be sub-let

The prohibition against sub-contracting the whole of the Works is primarily to prevent assignment by the 'back door', since in either case the practical effect on the project will be the same. While there is no express description of what is meant by 'the whole of the Works', it is arguable that a Contractor who provides minimal supervision and undertakes some or all of the temporary works, but sub-lets the whole of the permanent works is retaining part of the Works to himself and sub-contracting the rest, it would certainly be infringing the spirit of Clause 8.2, since he is sub-contracting the whole of that part of the Works which is of benefit to the Employer under the Contract. If such an arrangement is submitted to the Engineer it should thus be refused consent out of hand.

Engineer's consent

As has already been stated, in order to sub-contract any part of the Works the Contractor must first obtain the (preferably written) consent of the Engineer. The 'part' to be sub-contracted will usually mean either a geographical section of the Works or the work of one particular trade. Such consent is not to be 'unreasonably' withheld, and the Engineer may properly be called on to give reasons (again, preferably in writing) before he rejects any particular request for consent.

Upon what ground, then, might it be reasonable for the Engineer to refuse consent? Some have already been mentioned, but, in addition, should a proposed sub-contractor have performed badly in any particular way in connection with previous projects administered by the Engineer, or of which he has knowledge, the withholding of consent might well be justified so as to protect the Employer's interests. If special conditions have been bound into the Contract (such as the need to qualify under local by-laws, be on a specified list of approved contractors or the like) which the proposed sub-contractor does not meet, or if there is genuine doubt about his ability to show the specific skills or technical competence prescribed by or necessarily to be implied in the Specification, then it would be reasonable to withhold consent.

A particular area of difficulty is where the Engineer's earlier bad experience of a particular sub-contractor was really a problem of personalities. If the personnel involved in the earlier fracas (whether on the sub-contractor's or the Engineer's side) are to be engaged on the Works now to be carried out, it may well be sensible to foresee similar trouble and to allow this to affect the decision whether or not to give consent. In such cases written reasons for refusing consent should not be given (and oral reasons should be suitably reticent) lest persons concerned in the earlier troubles acquire a right to sue for defamation.

However, if different personnel are to be retained — in particular in the supervisory grades — it could well be that the earlier troubles will not recur and there might then be no good reason to suppose that the sub-contractor's work would fail to be entirely satisfactory. Unfortunately, little more can be said in such cases, other than that they need to be handled carefully and sensitively.

The Engineer must not delay giving his consent, or be tempted into asking questions when the answers to them will have no bearing upon the reasonableness of withholding consent. The only effect of such questions would be to delay the giving of consent, which could lead to the Contractor putting forward a claim for delay.

There is no specific period within which the Engineer has to decide, but it must be given within a reasonable period. This period should be sufficient for the Engineer to make enquires about the sub-contractor, to visit his works elsewhere, and, if necessary, to interview him. The period will also be affected by the Period for Completion of the Contract — the shorter the contract period the shorter the period that would be reasonable for the Engineer to give or withhold consent.

In order to keep this period as short as possible the Contractor should supply as much detail as possible about the prospective sub-contractor when he applies for consent.

However, the Engineer must always bear in mind when refusing consent that if such action is later adjudged to have been 'consent unreasonably withheld' then the Employer will be exposed to a claim by the Contractor for breach of contract. The measure of the damages would be the difference in the tendered price of the refused sub-contractor and that of the sub-contractor subsequently employed together with any additional costs that the Contractor might have been put to as a result of having to find another sub-contractor, such as delay costs whilst negotiations leading up to the sub-contract took place, and any additional facilities that may be required.

Effect of sub-letting without consent

If the Contractor does sub-contract any part of the Works without first obtaining the Engineer's consent, he will be in breach of the Contract. Yet there is no express provision in the minor works form setting out the consequences of such a breach or the Employer's remedy in such cases. Of course, the Employer can always have recourse to his remedies for breach under the common law, but even here the position is by no means absolutely clear.

If the circumstances of the breach are sufficiently serious, it may be that the breach might amount to a repudiation of the Contract, in which case the Employer would be entitled to treat the Contract as being at an end and expel both sub-contractor and Contractor alike. Normally, under most construction contracts the Employer is given an express right to terminate the Contract if the Contractor sub-lets without consent. But in *Thomas Feather & Co. (Bradford) Ltd* v *Keighley Corporation* (1953) 52 LGR 30 Lord Goddard appeared to hold that sub-contracting part of the Contract in breach of an express contract term did not amount to a repudiatory breach. This may not necessarily be so in general, since such cases tend to turn on their own particular facts. Nevertheless, sub-contracting without consent cannot now be said automatically to amount to repudiation even where an express power to terminate exists, so the Employer's right to terminate in such circumstances under the minor works form is even more in doubt.

Whether or not there is an express power to terminate, other remedies which at first sight might appear to be available, such as withholding payment for work done or refusing acceptance of completed work, cannot be exercised merely because the work has been carried out by a sub-contractor to whose employment the Engineer has not given consent, provided always that the work in question is otherwise in conformity with the Contract. However, although the Employer is bound to make the Site available to the Contractor and any of his sub-contractors to whom the Engineer has consented, it is arguable that the Employer could evict the unwanted sub-contractor from the Site. Nevertheless, such action would be fraught with difficulty in that the Contractor has possession of the Site during the currency of the Works and the Employer might be held to be guilty of trespass against the Contractor if he entered without the Contractor's permission.

In short, it is hard to see any way in practice for the Employer or Engineer to prevent the employment of an unwanted sub-contractor, other than to apply to the court for an injunction. Since the Engineer cannot withhold consent unreasonably, it would, it is suggested, be extremely difficult to persuade a court that to sub-let without permission amounts to a repudiatory breach, unless there are other supporting circumstances, such as an Engineer's express instruction forbidding sub-letting to a particular sub-contractor.

Nevertheless, no Contractor should sub-contract work lightly without the Engineer's consent, or even go ahead without bothering to ask for consent. Wilful disregard of Clause 8.2 might not in the event be deemed a repudiatory breach, but it would certainly lead to a breakdown of relationships with the Engineer, a breakdown from which the Contractor himself must ultimately suffer.

Contractor's responsibility

The Contractor remains fully responsible for the work of any sub-contractor and the consent of the Engineer does not affect his liability under the Contract in any way.

8.3. The Contractor shall be responsible for any acts defaults or neglects of any sub-contractor his agents servants or workmen in the execution of the Works or any part thereof as if they were the acts defaults or neglects of the Contractor.

Specialist works

Provision must be made in the Contract where it is not intended that the Contractor shall undertake particular items of the contract work himself. This may arise either because he is not thought to have the necessary expertise, or because the Employer (or the Engineer) wishes a particular specialist contractor be engaged to execute that part of the Works.

The Engineer has no power to order the Contractor to engage a certain specialist after the Contract has been awarded.

There are three particular matters concerning work of a specialist nature that need to be noted. They are as follows:

(i) the Contractor is not responsible for design of any part of the Works unless express provision is made for it, and
(ii) there is no provision for nominated sub-contractors, but
(iii) the naming of specialist sub-contractors in the Contract is permitted, subject to certain qualifications.

Commentary

Design responsibility

3.7. The Contractor shall not be responsible for the design of the Works, except where expressly stated in the Contract.

The words 'stated in the Contract' mean that unless the design obligation is expressly stated therein there is no means of imposing it upon the Contractor after the Contract has been let. An 'express statement' is one which is clear beyond doubt and requires no deduction to determine its meaning. Such statements should preferably be incorporated in the Specification, with full details of exactly which elements have to be designed, the performance or quality to which they are to be designed, and referring to the drawings upon which any layout requirements are shown.

No variation order issued by the Engineer can include any element of design responsibility unless the Contractor expressly agrees thereto. If the Engineer wishes design to be undertaken by persons other than himself after the Contract has been let then he must enter into a separate design contract with those persons.

Contractors should check all variation orders to be sure that they contain no design responsibility because, in addition to taking on liabilities that they have not contracted to accept, they will also find that their 'all-risks' insurance policies may not cover any failure due to design and may even be invalidated.

No nominated sub-contractors

The reason for the omission of provisions for nominated sub-contractors is that the minor works form is intended to be used for short period contracts where the design is complete prior to the issue of tender documents. The usual reason for needing a nominated sub-contractor, namely that the design is still continuing at the time of tender, will not, therefore, normally apply.

Specialist sub-contractors

1. Design

The Notes for guidance draw particular attention to contracts where design work of a specialist nature is required to be undertaken by a specialist sub-contractor or supplier (such as piling, structural steel, or electrical and mechanical installations).

NG6. If the Contractor is required to be responsible for design work of a specialist nature which would normally be undertaken by a specialist sub-contractor or supplier (such as structural steelwork, mechanical equipment or an electrical or plumbing installation) full details must be given in the Specification or in the Appendix to the Conditions of Contract or on the Drawings indicating precisely the Contractor's responsibility in respect of such work.

It is essential that this guidance is followed and that an exact definition of the boundaries of responsibility is spelt out in the contract documents because, unless this is done, the Contractor will have no responsibility for it. The Employer will then be liable for any loss should there be any failure in the design.

2. *Construction*

Where work is required to be executed by a named specialist sub-contractor, the name of that specialist sub-contractor and full details of what is required must be incorporated in the contract documents.

Provided that full details of the specification and conditions under which it is to be carried out are incorporated in the tender documents, the Contractor can, prior to submitting his tender, negotiate with the specialist and agree conditions of sub-contract and any particular terms. If the sub-contractor imposes any particular conditions upon the tendering contractor then that contractor can in turn either include a sum of money in his tender price to cover any additional risks that the sub-contractor imposes, or incorporate the same conditions in his tender to the Employer.

Thus, should the tendering contractor be unable to negotiate satisfactory terms, he can either refuse to tender, or qualify his tender to the extent necessary to meet the risk of employing the named specialist.

However, the naming of a particular *single* specialist should only be done in exceptional circumstances, such as where no one else would be acceptable for reasons of design or competence. Normally, a better alternative would be to list two or more acceptable sub-contractors from whom the tendering contractors may choose.

Note 7 in the Notes for guidance spells this out very clearly.

> NG7. The Engineer may in respect of any work that is to be sub-let or material purchased in connection with the Contract list in the Specification the names of approved sub-contractors or approved suppliers of material. Nothing however should prevent the Contractor carrying out such work himself if he so chooses or from using other sub-contractors or suppliers of his own choice provided their workmanship or product is satisfactory and equal to that from an approved sub-contractor or supplier.

The clear intention is that where the Engineer knows of suitable people for particular work, he should be free to put those names into the Contract, so that the Contractor has a chance to negotiate terms and conditions prior to submitting his tender. Note 7 also suggests that this should not be an exclusive process, but that the Contractor should be free to go to other people, or to do the work himself, if he so wishes. This is to ensure that the Employer is given the maximum benefit of competition.

Again, in the period between tendering and acceptance of the tender, the trading position of the sub-contractor, chosen from the list, may have changed significantly, or he may no longer be willing to undertake the work at all. The Contractor could thus be exposed to the risk, until his tender is accepted and he is able to conclude a sub-contract with the named specialist, that he will have to engage another specialist, presumably at a higher price. This risk must be reflected in his tender price. The more restricted his choice the greater will be the risk, and the price to the Employer.

If there is only one named specialist sub-contractor, the contractor should enter into an agreement with that sub-contractor to the effect that if the contractor is awarded the contract then the named specialist will be willing to enter into a contract with the contractor on the agreed terms.

Sub-contract documentation

There is no standard sub-contract document published for use with the minor works form. The FCEC Form of Sub-Contract ('the blue form') may be adapted for use with it, and details of the necessary amendments and guidance on how it should be completed are given in Chapter 17. These amendments have been discussed and agreed with the FCEC by the authors.

Comparison with ICE 5

There are two fundamental differences between the two contracts. First, with regard to assignment, under ICE 5 the Employer can assign at will, whereas under the minor works form he must obtain the consent of the Contractor. Under both contracts the Contractor has to obtain the consent of the Employer before any assignment can be made. Secondly, with regard to normal sub-letting under ICE 5, Clause 4, the Engineer has an absolute right to withhold permission to sub-let work, and the Contractor has then to execute the work himself, whereas under the minor works form the Engineer can only withhold his consent for good reason.

In the current operational state of the industry, the position under ICE 5 is becoming increasingly hard to justify, because the majority of all work on all major contracts is now sub-let. The provision under the minor works form is the more realistic in this respect. It also provides an administratively easier regime without detracting too much from the Employer's security.

There is also no provision in the minor works form for the Employer to terminate the Contract if the Works are sub-let 'to the detriment of good workmanship, or in defiance of the Engineer's express instructions to the contrary' (ICE 5, Clause 63). The Employer is left to his common law right of termination for repudiatory breach by the Contractor. It is doubtful if failure to obtain the consent of the Engineer would of itself amount to such breach.

There is no provision in the minor works form for prime cost items, or nominated sub-contractors similar to those in Clauses 58 and 59 of ICE 5. This is because it is intended that the design should be complete by the time tenders are invited. Where specialists are required to execute certain works the works in question and the specialist concerned must be named in the Contract and the Contractor invited to price the work on that basis. Provisional sums are permitted for contingencies.

The inclusion of the responsibility for design is permitted under both contracts, but only if it is expressly specified in the contract documents (ICE 5 Clause 8(2) and minor works 3.7). This means that the scope of the design responsibility must be fully spelt out in the documents for both contracts.

Comparison with ICE 6

On assignment ICE 6 follows the lead set by the minor works form and now neither party can make any assignment without the consent of the other. However, that consent must not unreasonably be withheld.

The Contractor is now free to sub-let any part of the Works without the consent of the Engineer. He must inform the Engineer of the name of the sub-contractor before that sub-contractor starts work or, in the case of a designer, on appointment. He still may not sub-let the whole of the Works.

This freedom is, however, counter-balanced by new powers given to the Engineer, under Clauses 4(5) and 16, to require the Contractor to remove from the Site either the sub-contractor himself or or any of the sub-contractor's employees if they are incompetent, misbehave or do anything detrimental to safety.

Although the provisions with regard to 'Nominated Sub-contractors' have been entirely re-drafted there is only a minor redistribution of risk between the parties.

Legal comment

Assignment

In common law the general rule of privity prevents a contract between A and B from conferring either rights or liabilities thereunder upon C. However, in commercial contracts at least, it is legally possible for either A or B to assign his *benefits* under the contract to C. If A assigns his benefits to C it will not usually be necessary for A to inform B that the benefits have been assigned, as there will be no contractual relationship between A and C and neither A nor C will be able to sue each other on the original contract. Nevertheless, A will be deemed to have agreed with C that the latter may if necessary sue B on the original contract using A's name.

The burdens of a party to the original contract can also be assigned, but normally only with the consent of the other party. Thus a purported assignment of burdens which is not at least brought to the notice of the other party to the original contract will be of no effect, although it might give the assignee default rights against the assignor. However, once notice has been given, consent may be implied from conduct and need not be given expressly or in writing. Should the assignee fail to discharge the assignor's duties under the original contract, however, the other party to that contract can usually only sue the assignor (who is not automatically discharged from the original contract by the fact of assignment) unless he has consented to the assignment, in which case he will usually be able to proceed against the assignee direct.

In short, and as stated in Chitty on Contracts (25th edition) at paragraph 1309:

> Everybody has a right to choose with whom he will contract and no one is obliged without his consent to accept the liability of a person other than him with whom he made his contract.

Novation

Assignment must not be confused with novation. Assignment is a transaction whereby a third party in effect takes over the role of one of the parties to an existing contract. However, the assignee never becomes a party to the original contract and the assignor never ceases to be a party to that contract. Novation, on the other hand, is a transaction by which, with the consent of all three parties concerned, a new contract is substituted for the original contract. Thus, instead of C's taking over A's function in a contract between A and B, novation means that A, B and C all mutually agree to rescind the contract between A and B and replace it with a new contract between C and B. It follows that the outgoing party A is never a party to the new contract (often somewhat inelegantly called the novated contract) and C is a full party to the new contract from the start. Putting it another way, there are in fact three contracts; the original one between A and B which is brought to an end; a contract of novation between A, B and C; and the new or novated contract between C and B.

Sub-contracting

Instead of assignment or novation, either party to the original contract can employ some third party to perform his part of the contract. This is known as vicarious performance and the principal remains fully liable to the other party to the original contract for the due performance of his obligations, and thus for the acts, omissions and defaults of his sub-contractor. In general, unless the obligations to be performed under the original contract are of a personal nature (which is not the case with construction contracts), the other party cannot object to the employment of a sub-contractor, provided that the benefits due under the contract are delivered in full and on time. However, if anything does go wrong, the injured party cannot sue the sub-contractor directly since there is no privity of contract between them, but must instead seek compensation from the sub-contractor's principal.

Nominated sub-contractors

All the ICE forms of contract contain express provisions defining and limiting the rights of the parties to assign or sub-contract. (Novation is not covered, since both original parties and the intended replacement party must expressly agree together.)

Unfortunately, ICE 5 (and ICE 6) contains provision for the Employer to nominate certain of the Contractor's sub-contractors. Such nomination also prevents the Contractor from electing to do that part of the Works himself. Thus the Contractor's freedom of choice in the matter is doubly restricted. It would appear that this pernicious development first grew up in the building industry and was later taken up by the Employers in civil engineering construction to increase their control over specialist or particularly sensitive parts of the Works.

This practice of nomination is anomalous in that it destroys the legal independence of the Contractor. This is important, since while the Contractor remains independent the Employer will not normally be liable to third parties for the consequences of any fault or misconduct by the Contractor. Conversely, if the Employer so intervenes in the Contractor's conduct of the Works that the Contractor effectively is obeying the Employer's orders, the Contractor becomes in law the Employer's agent, and his misdeeds towards third parties will be in law the Employer's misdeeds also.

For this reason, all forms of contract which contain a power to nominate sub-contractors also make more or less Draconian provision for the Contractor 'to save harmless and indemnify' the Employer against liabilities to third parties. The Employer is thus trying to have his cake and eat it. Indeed, a substantial proportion of construction disputes which eventually reach the courts probably arises from nomination in some form or another, and it is almost always extremely difficult to predict precisely what effect the courts will give to any particular set of circumstances.

It is therefore warmly to be welcomed that the draughtsmen of the minor works form have set their collective face against allowing the nomination of sub-contractors. It is to be hoped that the civil engineering industry will allow itself sooner or later to be weaned off this particular form of legal aberration.

Chapter 12

STATUTORY OBLIGATIONS

Introduction

The general requirement of the Contract is for the Employer to obtain those permissions that are necessary for the Works in their permanent form (that is when they have been constructed) and for the Contractor to obtain all permissions and permits to enable the Works to be executed.

Action by Contractor

The Contractor is required to give the various notices under the Public Health Acts, Building Regulations and other statutes, regulations and by-laws as may be necessary to enable the Works to be executed, and to meet all the costs involved in so doing.

> 9.1. The Contractor shall, subject to Clause 9.3 comply with and give all notices required by any statute statutory instrument rule or order or any regulation or by-law applicable to the construction of the Works (hereinafter called 'the statutory requirements') and shall pay all fees and charges which are payable in respect thereof.

Such matters will include:

- (i) rates on temporary accommodation, for both the Contractor and the Resident Engineer,
- (ii) scaffold licences,
- (iii) hoarding or gantry licences,
- (iv) consents to the temporary deposition of building materials or skips on highways,
- (v) licences to fell trees which are the subject of preservation orders,
- (vi) notices of commencement of work to both the Building Officer and the Factories Inspectorate,
- (vii) stage inspection notices,
- (viii) consents to open up highways and make connections to existing sewers,
- (ix) fire certificates in respect of using certain premises for offices, stores or workshops,
- (x) licences to store flammable liquids, and
- (xi) notices relating to employment, e.g. redundancy, employment of persons under 18, advising the ambulance service when more than 25 persons are employed.

The Contractor must allow for the costs of any fees or charges in connection with these matters in his tender price.

Action by Employer

9.2. The Employer shall be responsible for obtaining in due time any consent approval licence or permission but only to the extent that they may be necessary for the Works in their permanent form.

This requirement is strictly limited by the words 'necessary for the Works in their permanent form'. The Employer is therefore responsible for obtaining planning permissions and for ensuring that the Works comply with the Building Regulations. He must also obtain such statutory approvals as may be necessary under (for instance) Section 38 of the Highways Act 1980 in relation to the construction of roads, Sections 17 and 18 of the Public Health Act 1961 with regard to drainage, and must ensure compliance with the Control of Pollution Act 1974 when there is discharge of water into water courses.

The Employer must also obtain any licences or consents which may be necessary should the permanent works make use of materials or processes which are the subject of patents and, by necessary inference, pay any necessary licence fees or royalties which such use may entail, unless the Contract otherwise provides (either in the Specification or in the Bill of Quantities). Similarly, any problems about wayleaves or other private rights affecting the land upon which the permanent works are to be constructed are for the Employer to resolve.

However, where such consents, approvals, licences, permissions or wayleaves are necessary only for the Contractor's temporary works or methods of construction, they will normally be his responsibility.

Contractor not liable

9.3. The Contractor shall not be liable for any failure to comply with the statutory requirements where and to the extent that such failure results from the Contractor having carried out the Works in accordance with the Contract or with any instruction of the Engineer.

The Contractor has an implied duty in common law to bring to the attention of the Engineer (as the Employer's agent) any aspect of the design of the Works which in his opinion is in breach of any statute or regulation, or for which any required permission or consent has not been obtained. Nevertheless, Clause 9.3 expressly relieves the Contractor of any responsibility for such shortcomings, at least as between the Contractor and the Employer. It follows that, should the Contractor incur any penalty because the design is in breach of some statutory requirement, he is entitled to pass the cost on to the Employer.

New Roads and Street Works Act 1991

ICE 5, Clause 27, contains special provisions covering the notices, procedures and other special requirements of the Public Utility Street Works Act 1950 (PUSWA), but it was not thought necessary to include equivalent coverage in the minor works form. However, this could have led to difficulty, since the provisions of Clauses 9.1, 9.2 and 9.3 might in some circumstances have been less than adequate with regard to PUSWA.

Fortunately, this lacuna no longer exists, since PUSWA has now itself been repealed and replaced by the New Roads and Street Works Act 1991. While it is as yet too soon to be quite sure, it would appear that Clauses 9.1, 9.2 and 9.3 now cover these matters adequately. Nevertheless, at least for the time being and to avoid unnecessary delays, it would seem sensible for the Engineer specifically to draw the Contractor's attention to the need for serving any notices under the new Act, or even for the Engineer to serve such notices himself. At the very least, matters concerning the new Act should be considered at the first progress meeting following the award of the Contract, and they could with advantage be the subject of discussion at the pre-Contract stage. In certain cases it may be prudent to include such matters in the Specification.

Comparison with ICE 5

These clauses cover the same ground as Clause 26 in ICE 5. The provisions of the minor works form are much simpler but the intention of both contracts is broadly the same. The only exception is that the Contractor is expected to include in his tender price the cost of 'all fees and charges' under the minor works form but under ICE 5 they are to be repaid by the Employer. This difference is surprising, since a small Contractor for minor works may well know less about such matters than a Contractor executing work under ICE 5.

As stated above, the minor works form did not deal with the special requirements of PUSWA. Matters under the new Act are ordered differently and ICE 5, Clause 27, is itself now largely out of date.

It should also be noted that under ICE 5, Clause 42(2) the Contractor is required to organize access to the Site and pay any costs that may be involved in providing such access. Under the minor works form access is the sole responsibility of the Employer (Clause 4.4 (e)).

Unlike ICE 5, Clause 28(2), the minor works form does not deal with patent rights or royalties. Since the design should normally be complete before tenders are invited, it is logical that the Employer (or the Engineer) should ensure that the permanent works contravene no patent rights or copyright.

Finally, there are no minor works equivalents to the following clauses of ICE 5:

Clause 29(1): Interference with traffic and adjoining properties. This is not a serious omission, since the Contractor will be liable in law for any interference with traffic (unless it is the unavoidable result of the carrying out of the Works, in which case Clause 10.4 makes the Employer responsible) and damage to adjacent property will be covered by the Contractor's indemnity under Clause 10.2.

Clause 29(2): Noise and disturbance. Again, the omission of this provision would not seem significant. The Employer is unlikely to be under vicarious liability for noise and disturbance caused by the operations of the Contractor, which in itself is unlikely to cause damage to persons or property. Such disturbance, therefore, will not be covered by Clause 10.4(d), and the Contractor will be wholly responsible for claims arising from such causes.

Clause 30(1): Avoidance of damage to highways, etc. Broadly, this clause requires the Contractor to indemnify the Employer against any damage to highways or bridges caused as a result of the carrying out of the Works, except when the damage is caused by transporting materials or manufactured or fabricated articles for the execution of the Works. It therefore separates that which is within the Contractor's control from that which is outside it. Where the damage is caused by delivery of materials specified by the Engineer, then the Employer may be responsible for damage occasioned to highways.[1]

In the minor works form damage to highways is covered by the indemnity given by the Contractor to the Employer in Clause 10(2), and will be qualified by the counter-indemnity given to the Contractor by the Employer under Clause 10.4(d) for damage which is the unavoidable result of the construction of the Works. This matter is discussed further in Chapter 13.

Comparison with ICE 6

When ICE 6 first appeared, PUSWA was still in force, so Clause 27 was included exactly as in ICE 5. Since PUSWA has now been replaced by the New Roads and Street Works Act 1991, Clause 27 in its old form is no longer needed, and it is understood that ICE 6 will shortly be amended by deleting Clause 27 and including a suitable reference to the new Act in Clause 26(3). That apart, there is no material difference between ICE 5 and ICE 6 in these matters, except that the word 'pollution' has now been added to Clause 29(2). The foregoing comments will apply.

1. But see the New Roads and Street Works Act 1991.

Chapter 13

LIABILITIES AND INSURANCE

Introduction

The Conditions of Contract allocate the risks of construction between the parties. A contract of insurance is one whereby an insurance company indemnifies the party taking out the insurance policy for damage suffered as a result of the incidence of one or more of these risks. Some risks are not insurable at all (see legal comment at the end of this chapter). Of those which can be insured, some are expressly to be borne by the Employer and are included in the Excepted Risks (see below).

Most construction contracts require that the Contractor shall take out insurance to cover

(i) risks of damage to the Works,
(ii) risks of damage to persons or property resulting from the execution of the Works (third party risks), and
(iii) damage to persons involved in the Works (Employer's Liability).

Damage to the Works

Damage to the Works is usually covered by a policy known as Contractor's all risks (CAR), which will generally cover loss or damage to

(i) the Contract Works, consisting of the permanent and temporary works to be constructed in the performance of the Contract, together with all the materials for use in connection therewith;
(ii) tools, plant, equipment and temporary buildings, the property of the insured and/or for which he is responsible; and
(iii) personal effects and tools of the insured's employees insofar as these are not otherwise covered.

The policy, however, will not cover loss or damage resulting from the Excepted Risks, or any uninsurable risks.

The policy will be valid only for so long as the Contractor has responsibility for these matters, and this generally ceases within a short period after the completion of the Works. The Contractor will often take out a CAR policy to cover the liabilities of the Employer, the Contractor and all of his sub-contractors. All are then termed the 'joint insured'. This has the advantage of ensuring that any damage that occurs to the Works is covered by one policy, thereby diminishing the scope for arguments between different insurance companies as to which of them is liable.

Damage to third parties

The second type of policy which the Contractor is required to take out is a public liability (or third party) policy. This will provide indemnity against personal injury claims made by the public (but not employees) and property damage claims by any third party.

It is important to the Employer that such policies should be in force at all times during the construction of the Works, since any third party who suffers damage as a result of the execution of the Works may have a right of action against the Employer as the instigator of the Works. Thus, for instance, if a neighbouring property should suffer subsidence as a result of excavation on an adjacent site, the owner of that property could take action against the Employer for the damage that he has suffered.

Employer's liability

The third type of insurance covers the general liability at law of any employer (in this case, the Contractor) to his employees for bodily injury or death arising out of or in the course of their employment and is usually called Employer's liability insurance.

Being a requirement imposed by statute, it is not possible to contract out of this liability and it is unlawful not to take out appropriate insurance. Thus, at first sight, there would seem to be no need for this class of insurance to be required by the Contract as well. However, should the Contractor become insolvent and be without the appropriate insurance, the statutory liability would fall upon the Employer; hence the inclusion in the Contract that the Contractor shall insure. By making it a contractual requirement as well as a statutory one, the Employer is entitled to check that the Contractor has actually taken out and kept in force such insurance cover.

Under the Employer's Liability (Compulsory Insurance) Act 1969 such insurance must be taken out and maintained for an amount of not less than £2 million in respect of claims arising out of any one occurrence and relating to one or more employees. It is usual, however, for the amount of such insurance to be unlimited. In any event, such insurance will normally be taken out on an annual basis to cover all work in hand by the Contractor, and will not usually be limited to any one project.

General

All contracts of insurance are based on the principle of 'utmost good faith' (*uberrimae fidei* — see legal comment at the end of this chapter). This means that the party seeking to take out insurance must bring all the facts and other information which are, or may be, relevant to the risk to be covered to the attention of the insurer when filling out the proposal form.

Thereafter, the insurer must be informed as soon as possible of any change in circumstances which might affect the risks covered. Thus no changes in the Conditions of Contract or the Specification should normally be agreed to without first consulting the insurer. Similarly, variations which involve work of a character different from that originally envisaged should be reported to the insurer even though the Contractor is contractually bound to carry them out.

Finally, insurers should be notified as soon as possible when a possible claim situation arises, so that the insurer, if he so wishes, can be a party to decisions affecting remedial action and costs. Failure to keep the insurer fully informed at all times can result in the policy's being repudiated by the insurer, leaving the policy-holders without cover.

Insurances against damage to the Works

10.1. (1) If so stated in the Appendix the Contractor shall maintain insurance in the joint names of the Employer and the Contractor in respect of the Permanent Works and the Temporary Works (including for the purpose of this Clause any unfixed materials or other things delivered to the Site for incorporation therein) to their full value and the constructional plant to its full value against all loss or damage from whatever cause arising (other than the Excepted Risks) for which he is responsible under the terms of the Contract.

Insurance cover against damage to the Works (usually CAR, see above) is optional. It is thus open to the Employer to take out this cover himself. For instance, on a scheme involving a number of small contracts all let on the minor works form it might be more convenient (and could prove cheaper) for the Employer to insure the whole scheme instead of relying on each contractor to take out a small, independent policy. This would also have the advantage of minimizing disputes on liability should damage occur, since only one policy would be involved. For similar reasons, large contractors often take out CAR cover on behalf of themselves and all of their sub-contractors either on an annual basis for all work in progress or on a project-by-project basis.

The terms and conditions of insurance policies under the minor works form will normally be the same as under ICE 5, since many contractors will have a mixture of contracts involving both forms. Different insurance policies for different forms of contract could give rise to considerable practical difficulties and would not make commercial good sense.

If so stated in the Appendix . . .

The option of requiring the Contractor to insure the Works is incorporated into the Contract by deleting the words 'not required' from Item 15 (p. 14) of the Appendix to the Conditions of Contract. This will normally be done by the Engineer before issuing the tender documents. Failure to make this deletion would appear to mean that the Contractor is not required to take out insurance for damage to the Works. Engineers should therefore check Item 15 in the Appendix particularly carefully before inviting tenders.

. . . the Contractor shall maintain insurance in the joint names of the Employer and the Contractor . . .

The importance of insuring in the joint names of the parties to the Contract (who are then referred to as the 'joint insured') is that the insurance cover will then be available for loss of damage regardless of which party actually suffers the loss. Squabbles between separate insurance companies about which of them should meet any particular claim should thus largely be avoided.

... in respect of the Permanent Works and the Temporary Works (including for the purpose of this Clause any unfixed materials or other things delivered to the Site for incorporation therein) to their full value ...

The cover required for care of the Works is to the full value of the Works, and this will include any fees for design or other consultancy services that may be required in relation to it.

... and the constructional plant to its full value ...

This provision is a hang-over from the requirements in ICE 5, under which ownership of the Contractor's plant and equipment is vested in the Employer. That means that whilst the plant and equipment is on the Site the Employer is legally its owner, and if it is damaged there is a chance that the Employer might be liable for such damage.

Under the minor works form there is no provision for vesting, and therefore this provision is not strictly necessary. The requirement for the Contractor to insure his own plant and equipment has now been deleted from ICE 6 so this insurance cover is no longer required in order to achieve compatability of insurance policies for both forms of contract.

If the Contractor is not required to insure the Works under Clause 10(1) he will have to take out separate insurance cover for his plant and equipment.

... (other than the Excepted Risks) ...

The Excepted Risks are defined in Clause 1.5. It should be noted that included in the Excepted Risks are loss or damage resulting from 'a cause due to the use or occupation by the Employer, his agents, servants or other contractors (not being employed by the Contractor) of any part of the Permanent Works'. It is therefore important that as soon as the Employer takes possession of the Works, or has any work done by his own workmen, he has the necessary insurance cover for damage to those parts of the Works.

Also in the Excepted Risks is loss or damage due to 'fault defect or error or omission in the design of the Works'. Normally this will be covered by the Engineer's professional indemnity policy (PI). If, however, the Employer is undertaking the design in-house he will need to take out professional indemnity cover himself.

Under the minor works form the Contractor cannot be instructed to undertake any element of design other than that included in the Contract. In any event, the Contractor's CAR policy will not normally cover any design of the Permanent Works which is undertaken by the Contractor. He will usually need to take out a separate PI policy to cover such work. However, sometimes insurers will agree to cover the design to the extent that it is included in the Contract at the time of formation.

Contractors must take care not to accept any variations which include any design element which is not defined in the Contract if their insurance policies are to remain valid.

These two Excepted Risks — use or occupation of the Works, and design of the Works — are the reasons why the Employer may find it more convenient to take out insurance cover for the care of the Works himself, and the minor works form provides that option. When he does so, Note 13(7) in the Notes for guidance point out that details of such insurance and any excesses which the Contractor may be expected to carry should be stated in the tender documents.

Excesses

NG13(7). *Clause 10.1 (Insurance of the Works).* This is at the option of the Employer. It must be borne in mind that Contractors frequently carry large excesses on their all-risks policies so that the Contractor then accepts the risk under Clause 10.1 in respect of any uninsured loss. When the insurance under Clause 10.1 is to be provided by the Employer the details of such insurance, including any excesses which the Contractor may be expected to carry should be stated in the tender documents.

'Excesses' are the amounts that are deducted by the insurance company from any claim prior to payment. It is important that the Employer, when negotiating insurance cover, should try to maintain the value of excesses normal in the contracting industry. To set the level of excess too high will encourage contractors to take out insurance cover for these alone, which will negate the benefits to the Employer of taking out the cover himself.

It is possible for the Contractor to comply with the Contract's requirements for insurance, but to accept excesses of a magnitude that would leave the Employer unreasonably exposed if the Contractor was to go into liquidation. The Engineer, when checking the Contractor's insurance policies, should therefore check the excesses and, if they are unacceptably large, try to persuade the Contractor to agree to have them lowered. He could give instructions to this effect under Clause 2.3(g) but if he did so the additional premium payable by the Contractor would be recoverable from the Employer under Clause 6.1. Alternatively, it would be open to the Employer to insure the excesses himself on a contingency basis. To avoid the problem the tendering contractors could be instructed in the instructions to tenderers to provide information on their proposed excesses with their tender.

Manner and period of insurance

10.1. (2) Such insurance shall be effected in such a manner that the Employer and the Contractor are covered for the period stipulated in Clause 3.2, and are also covered for loss or damage arising during the Defects Correction Period from such cause occurring prior to the commencement of the Defects Correction Period and for any loss or damage occasioned by the Contractor in the course of any operation carried out by him for the purpose of complying with his obligations under Clauses 4.7 and 5.2.

The Contractor is required to insure

(i) the whole of the Works during construction, from the starting date (Clause 4.1) until 14 days after the issue of a certificate of completion (Clause 3.2);

(ii) any work outstanding at practical completion (Clause 4.5) while it is being carried out, and rectification of any defects discovered before practical completion (Clause 4.7) until the issue of an Engineer's certificate under Clause 5.4 (i.e. during the Defects Correction Period); and

(iii) the rectification of any defects which may be discovered during the Defects Correction Period.

Such insurance shall be effected in such a manner that the Employer and the Contractor are covered . . .

This refers back to the requirement in Clause 10.1(1) that insurance shall be maintained in the joint names of the parties (see above).

. . . for the period stipulated in Clause 3.2 . . .

This period runs from the starting date (Clause 4.1) until 14 days after the Engineer issues the Certificate of Completion for the whole of the Works. This is not necessarily the day upon which the Works actually reach completion and which the Engineer certifies under Clause 4.5(3). It is therefore important for the Contractor to obtain the Certificate of Completion as early as possible after the completion of the Works, and pursue it with vigour if the Engineer is slow in issuing it.

The issue of the Certificate of Competion is therefore the means by which the Employer is reminded that the Contractor is no longer responsible for the care of the works and that he has to take out any necessary replacement cover within 14 days.

Contractor need not insure defective work

10.1. (3) The Contractor shall not be liable to insure against the necessity for the repair or reconstruction of any work constructed with materials or workmanship not in accordance with the requirements of the Contract.

The Contractor does not have to insure for the costs of replacing defective work. Defective parts or materials will generally be covered by a warranty from the supplier. Indeed, this may be something for which cover is not available at all, but if it is it will be a different type of cover called product liability insurance.

However, damage resulting from the defective work or materials will be covered by the CAR policy. For instance, if a bridge bearing fails due to faulty workmanship and the bridge collapses, the consequential cost of rebuilding the bridge would be covered by the CAR but not the replacement of the bearing that failed.

Third party insurance

Clauses 10.2 to 10.5 allocate between the Employer and the Contractor the risks of damage to third parties resulting from the execution of the Works.

10.2. The Contractor shall indemnify and keep the Employer indemnified against all losses and claims for injury or damage to any person or property whatsoever (save for the matters for which the Contractor is responsible under Clause 3.2) which may arise out of or in consequence of the Works and against all claims demands proceedings damages costs charges and expenses whatsoever in respect thereof or in relation thereto subject to Clauses 10.3 and 10.4.

In general, the Contractor indemnifies the Employer against all third party risks, but this general indemnity is reduced so as not to include

(i) acts of the Employer and those acting under him,
(ii) those damages which are the inevitable result of the Works, and
(iii) damage inherent in occupation of the Site.

Clause 10.2 is the general indemnity given by the Contractor to the Employer in relation to all claims by third parties as a result of the Works. This indemnity is reduced in accordance with the requirements of Clauses 10.3 and 10.4.

10.3. The liability of the Contractor to indemnify the Employer under Clause 10.2 shall be reduced proportionately to the extent that the act or neglect of the Engineer or the Employer his servants or agents or other contractors not employed by the Contractor may have contributed to the said loss injury or damage.

This clause reduces the liability of the Contractor in proportion to the extent that the Employer or those employed by him are to blame for the cause of the damage. No mechanism is prescribed for assessing the relative contributions of the parties, and no discussion should be entered into between the Engineer and Contractor concerning such contributions without the agreement of the insurer.

10.4. The Contractor shall not be liable for or in respect of or to indemnify the Employer against any compensation or damage for or with respect to:-

(a) damage to crops being on the Site (save in so far as possession has not been given to the Contractor);
(b) the use or occupation of land (which has been provided by the Employer) by the Works or any part thereof or for the purpose of constructing completing and maintaining the Works (including consequent losses of crops) or interference whether temporary or permanent with any right of way light air or water or other easement or quasi easement which are the unavoidable result of the construction of the Works in accordance with the Contract;
(c) the right of the Employer to construct the Works or any part thereof on over under in or through any land;
(d) damage which is the unavoidable result of the construction of the Works in accordance with the Contract;
(e) injuries or damage to persons or property resulting from any act or neglect or breach of statutory duty done or committed by the Engineer or the Employer his agents servants or other contractors (not being employed by the Contractor) or for or in respect of any claims demands proceedings damages costs charges and expenses in respect thereof or in relation thereto.

These are all matters which are peculiarly within the control of the Employer or which necessarily arise from the very existence of the permanent works. It is thus both right and just that the Contractor be relieved of the burden of responsibility for such matters.

10.4 (a) damage to crops being on the Site (save in so far as possession has not been given to the Contractor).

The Contractor is under no obligation once he is given possession of the Site to check whether crops can be removed.

10.4 (b) the use or occupation of land . . .

The Contractor is entitled to assume that there is no legal or other impediment preventing him from entering upon the Site and carrying out the work in accordance with the Drawings and Specification. If there is, the consequences will be the Employer's responsibility (see Clause 10.5 below).

To safeguard the Employer's position the Engineer must ensure that no such impediments exist in relation to the Site or in connection with any access to the Site. If this proves not to be possible, any restrictions which cannot be lifted or avoided must be set out in the invitation to tender and expressly included in the contract documents.

> 10.4 (c) the right of the Employer to construct the Works or any part thereof on over under or through any land;

The Employer must ensure that the Works have planning permission and are in accordance with the Building Regulations. The Contractor has a statutory duty to comply with Building Regulations and should point out any deviations therefrom of which he becomes aware. The Employer must also ensure that there are no other restrictions such as covenants on the land, or other limitations on its use.

> 10.4 (d) damage which is the unavoidable result of the construction of the Works in accordance with the Contract;

This covers the general principle that the Contractor undertakes works as directed by the Engineer and has no duty to do other than execute the Works in accordance with the Contract and the law.

> 10.4 (e) injury or damage to persons or property resulting from any act or neglect or breach of statutory duty done or committed by the Engineer or the Employer . . .

The Engineer should note that he is not covered for the results of his own actions under the policy to be taken out by the Contractor.

Employer to indemnify the Contractor

> 10.5. The Employer will save harmless and indemnify the Contractor from and against all claims demands proceedings damages costs charges and expenses in respect of matters referred to in Clause 10.4. Provided always that the Employer's liability to indemnify the Contractor under paragraph (e) of Clause 10.4 shall be reduced proportionately to the extent that the act or neglect of the Contractor or his sub-contractors servants or agents may have contributed to the said injury or damage.

This clause is the counterpart to Clause 10.3, and the comments under that clause apply equally to this one.

Employer to approve insurance

> 10.6. The Contractor shall throughout the execution of the Works maintain insurance against damage loss or injury for which he is liable under Clause 10.2. subject to the exceptions provided by Clauses 10.3. and 10.4. Such insurance shall be effected with an insurer and in terms approved by the Employer (which approval shall not be unreasonably withheld) for at least the amount stated in the Appendix. The terms of such insurance shall include a provision whereby in the event of any claim in respect of which the Contractor would be entitled to receive indemnity under the policy being

brought or made against the Employer the insurer will indemnify the Employer against any such claim and any costs charges and expenses in respect thereof.

The words 'throughout the execution of the work' do not define when such insurance is to come to an end. The intention seems to be that the Employer should take out third party insurance at the same time as he takes over the Works, or within 14 days of the issue of the Certificate of Completion, whichever is the earlier event.

. . . subject to the exceptions provided by Clause 10.3. and 10.4.

The Employer will not be insured under the Contractor's policy for these matters, and needs to take out a separate policy for them where appropriate.

Such insurance should be effected with an insurer and in terms approved by the Employer . . .

There will be comparatively little time for such approval to be given. The Engineer should therefore consider asking Contractors to submit with their tender the names of the insurers whom they propose to use. Most contractors take out an annual public liability policy, and this may be open to inspection, but if they do not, reliance will have to be placed upon the Contractor's broker for confirmation of the wording, because policies take from three to four months to issue, and are unlikely to be available during the operative part of the Contract.

The Engineer should check the excesses and if necessary instruct the Contractor to arrange for the reduction of any excesses in his insurance policy which may be unacceptable to the Employer (see page 128).

. . . (which approval shall not be unreasonably withheld) . . .

Where the Contractor has an annual policy for all of his works there will be a presumption that it would be unreasonable not to accept the policy, provided that it satisfied the requirement as to amount.

. . . for at least the amount stated in the Appendix.

The amount of the insurance required is given in Item 16 of the Appendix. A simple figure is required to be inserted by the Engineer prior to dispatch of the tender documents.

The line 'any one accident/number of accidents unlimited' is a statement of requirement not an option. Neither provision should be deleted as they both apply. Note 13(8) in the Notes for guidance gives guidance on the amount required.

NG13(8). *Clause 10.6 (Third party insurance).* A minimum cover of £500 000 for any one accident/unlimited number of accidents should normally be insisted upon. In certain locations where there is greater risk to adjacent properties a higher limit may be desirable.

Works adjacent to railway lines or on airports generally attract a higher requirement. The Engineer in consultation with the Employer should agree what figure is required. The figure inserted in the Appendix does not limit the liability of the Contractor but may well be the limit of his ability to pay should a serious accident occur.

The terms of such insurance shall include a provision whereby in the event of any claim in respect of which the Contractor would be entitled to receive indemnity under the policy being brought or made against the Employer, the insurer will indemnify the Employer against any such claim and any costs, charges and expenses in respect thereof.

This final sentence of the clause is what is known as 'indemnity to principals' provision. This gives the Employer a right under the policy to claim directly against the insurers should a third party choose to sue him instead of the Contractor. In this it differs from the 'joint insured' provision of Clause 10.1(2) in respect of the CAR policy. The reason for this difference is discussed in the legal comment at the end of this chapter.

It should also be noted that under Clause 10.2 the Contractor is required to indemnify the Employer against all losses to 'person or property'. Such a person could well be an employee of the Employer who is not engaged in any way with the Works, and such property could include property of the Employer other than the Works. It follows that loss or damage to such persons or property would fall outside the scope of CAR insurance unless the normal policy was substantially modified, since it properly falls within the scope of third party cover.

Contractor to comply with terms of any policy

10.7. The Contractor shall comply with the terms of any policy issued in connection with the Contract and shall whenever required produce to the Employer the policy or policies of insurance and the receipts for the payments of the current premiums.

This is a sensible provision, but may not be feasible in practice. Receipts for the premium are usually available, as are cover notes, but the policies themselves can take three or four months to issue. The Engineer should request evidence of the payment of premiums and in the absence of the policy, get confirmation from the Contractor's brokers that the required policies are in force and of the date that they expire or, alternatively, when renewal premiums are payable.

Failure to issue

It is important to check that insurances are in place because until a claim has arisen and it can be shown that neither the Contractor's own funds nor insurance funds are available to meet it, the Employer is not in a position to show that he has suffered any damage. By that time it will usually be too late for him to have an effective remedy, since the claim will then become just another unsecured claim against an insolvent Contractor.[1]

Should the Contractor fail to insure as required under the Contract (or be unable to prove that he has done so), a prudent Employer should himself take out the necessary cover, having first given the Contractor due notice

1. See *The ICE Conditions of Contract, fifth edition: a commentary* by I. N. Duncan Wallace (1978), pp. 64–85. London: Sweet and Maxwell; and *Construction insurance* by Nael G. Bunni (1986), pp. 176–212. Barking: Elsevier Applied Science.

in writing that he intends to do so if satisfactory proof is not forthcoming by some suitable specified date.

It is for the Engineer in the first instance to check that the Contractor has insured, and to advise the Employer on appropriate action should the position appear in any way to be unsatisfactory.

While there is no express provision in the minor works form covering failure to insure, such failure is clearly a breach of contract entitling the Employer in common law to recover the cost of appropriate insurance by way of damages.

Comparison with ICE 5

The minor works insurance clauses have been drafted to be compatible with ICE 5, Clauses 20 to 25. Annual policies taken out for work under those general conditions will thus be equally applicable to minor works projects. There are, however, two omissions in the minor works form in that neither Clause 24 (Accidents or injury to workmen) nor Clause 25 (Remedy on contractor's failure to insure) of ICE 5 have any express equivalent in minor works form Clause 10. However, nothing is lost thereby, since workmen's compensation insurance is a statutory requirement which cannot be avoided and the remedy for failure to insure is an action for breach of contract.

Comparison with ICE 6

While there are some differences between the insurance requirements of ICE 5 and ICE 6 (mainly in the deletion of risks which are inherently uninsurable), the above comparison between the minor works form and ICE 5 remains equally valid for ICE 6.

Legal comment

An insurance policy is a contract under which the insurer undertakes to compensate the insured should any of the events specified in the policy come about so as to involve the insured in loss or damage, and the insured undertakes to pay the premium and to observe all the conditions set out in the policy.

Uberrimae fidei

Contracts of insurance are unusual in that the insured must make full disclosure of all circumstances which could possibly prove to be relevant to the assessment of the risk to be covered by the insurer. For this reason such contracts are said to be *uberrimae fidei* (i.e. matters of utmost good faith). Failure to make such full disclosure entitles the insurer to treat the contract as being of no effect, and to withhold benefit while retaining the premium.

Everything must be disclosed, not just that which the insured (at this stage, more correctly called the proposer) may think relevant.

In *Lindenau* v *Desborough* (1828) Bailey J. said

> I think that in all cases of insurance, whether on ships, houses, or lives, the underwriter should be informed of every material circumstance within the knowledge of the assured: and that the proper question is, whether any particular circumstance was in fact material? and not whether the party believed it to be so. The contrary doctrine would lead to frequent suppression of information, and it would often be extremely difficult to show that the party neglecting to give the information thought it material. But if it be held that all material facts must be disclosed, it will be in the interest of the assured to make a full and fair disclosure of all the information within their reach.

The extent of the duty to disclose is thus to declare everything that would influence the judgement of a prudent insurer in fixing the premium or in determining whether to take the risk at all. The proposer, however, cannot disclose facts of which he has no knowledge and the duty is therefore limited to those facts which he knows or ought to know.

This general duty can, however, be extended by express contract terms which may require the assured to agree to the accuracy of the information that he has provided. Accordingly it is not sufficient just to answer the questions on the proposal form. If there are other material facts within the proposer's knowledge they too must be disclosed [*Hair* v *Prudential Assurance Co. Ltd.* (1982)].

The insurance contract is only *uberrimae fidei* before the policy comes into effect. Thereafter it is only subject to the ordinary requirements of good faith. Nevertheless, some insurance contracts set out by means of express terms to reinstate the higher standard by making full disclosure of 'a condition precedent' (that is a strict requirement) to their making any payment under the policy.

Insurable risk

The insured must also have what is known as an insurable risk in the subject matter of the contract of insurance. This means that he must bear some relationship recognized by law with the subject matter so that if it is destroyed or damaged or he has to pay out as a result of some legal liability he will also suffer financially as a consequence. Thus a husband and wife can insure each other's life but neither could insure the life of their favourite television star if the only loss they would suffer as a result of the death of that person would be the loss of enjoyment of his or her performance.

Subrogation

Many insurance contracts are contracts of indemnity. If the insured risk occurs, the insurer pays the insured such compensation as will — so far as money can so do — restore him to the position he would have enjoyed had the risk not come to pass. Such liabilities may often arise from claims against the insured from third parties.

The insurer will naturally seek to exercise some control over the amount of such compensation he is called on to pay, and to that end will wish to negotiate with the claimant. But in the case of third party claims the insured is unlikely to have any contractual relationship with the third party claimant and at first sight, due to the doctrine of privity, there would appear to be no way in which the insurer can insist on negotiating with the claimant.

For this reason, most insurance contracts include a right of subrogation.

This enables the insurer effectively to step into the shoes of the insured and to negotiate (and, if necessary, litigate or arbitrate) directly with the claimant in the insured's name. A right of subrogation can also exist at common law, but can only be exercised before any payment is made. In such cases, negotiations are then taken completely out of the hands of the insured.

'Joint insured' and 'indemnity to principals'

The 'joint insured' provision under Clause 10.1(2) in respect of the CAR policy allows either party to claim under the policy for loss or damage to the permanent or temporary works, unfixed materials and construction plant. Thus, whatever the allocation of risk between the parties may be, each of them has an actual, or potential, interest in the subject matter of the policy.

CAR policies are concerned exclusively with loss or damage to the permanent or temporary works, unfixed materials and constructional plant, in all of which the Contractor and the Employer both have an actual or potential interest. Thus, should circumstances arise whereby one party is reluctant (or unable) to claim under the policy, it makes sense that the other party should be able to do so. It is for this reason that Clause 10.1(2) contains a 'joint insured' provision.

On the other hand, it will usually be the case that neither the Contractor nor the Employer will have any direct interest in the subject matter of a third party claim, and the claimant will be free to proceed against either of them, regardless of which of them is actually at fault. Difficulties could arise if, for example, the Employer was sued for something which he thought was the fault of the Contractor but the latter did not agree. In such circumstances the 'indemnity to principals' provision in Clause 10.6 allows the Employer to claim directly against the insurer, and also avoids the difficulty under joint insured arrangements that failure by one joint insured to observe all the conditions of the policy could disqualify both joint insured from claiming.

Chapter 14

DISPUTES

Introduction

From ancient times arbitration has been a well-established procedure for resolving commercial disputes without recourse to the courts, and some of the earliest forms of contract for civil engineering works contained arbitration clauses. Indeed, during the nineteenth century the Engineer appointed under the Contract to design and administer the project was often also appointed as sole arbitrator in any ensuing disputes, even though he might then be acting as judge and jury in a matter arising out of his own actions.

Needless to say, such Engineer arbitrators fell out of fashion and, certainly since the passing of the Arbitration Act 1934, it has generally been held that the Engineer cannot also be appointed arbitrator in connection with the same Contract or Works (unless, of course, the parties so agree in writing — but this almost never happens).

An echo of the nineteenth century practice still remained, however, in the first edition of the ICE Conditions of Contract in that a condition precedent to arbitration was inserted making it necessary to refer back to the Engineer any dispute or difference for his formal decision before either party could refer that matter to arbitration. This is still the case today, the condition precedent appearing in full force in Clause 66 of both ICE 5 and ICE 6.

The continued presence of this provision is justified in practice by the belief (hard to prove, but easy to experience) that such reference back is effective in settling the matter in dispute or providing a basis for the parties themselves to reach a compromise, thereby avoiding the need for arbitration in, perhaps, 90% of cases. Experience indicates that this is so even where the dispute arises out of some earlier act or decision of the Engineer, since it provides an opportunity for him to think again and, perhaps, refine his earlier decision or even vary it as a result of further information coming to light since the earlier decision.

This condition precedent has been omitted from the minor works form but, in its place, a new (and optional) procedure for conciliation has been introduced which, it is hoped, will preserve this very necessary contractual safety valve. This is particularly important for minor works projects, since full formal arbitration can be both protracted and expensive unless it is kept very firmly under control.

Notice of Dispute

11.1. If any dispute or difference of any kind whatsoever shall arise between the Employer and the Contractor in connection with or arising out of the Contract or the carrying out of the Works (excluding the dispute under Clause 7.9 but including the dispute as to any act or omission of the Engineer) whether arising during the progress of the Works or after their completion it shall be settled in accordance with the following provisions.

This is clearly based on the definition of 'disputes or differences' which can be referred to arbitration under Clause 66 of ICE 5, as set out in Sub-clause (1) of that clause.

The range of such 'disputes or differences' is very wide indeed, covering both problems with the Works themselves, or the carrying out of those Works, and the Contract itself as a legal entity. Thus matters concerning the formation of the Contract and other matters such as tort[1] or fraud, are all included.

. . . including . . . any act or omission of the Engineer . . .

This will cover decisions, opinions, instructions, valuations or certificates, together with any failure on the part of the Engineer to give or make such decisions, opinions, instructions, valuations or certificates if in the circumstances he should have acted. It will also, of course, cover any undue delay in taking such action, provided that such delay results in loss or damage to the Contractor (or, indeed, to the Employer — this clause is completely even-handed and both parties have equal rights to complain).

The Engineer should therefore always keep sufficient records both of the facts and of the reasons for any action or inaction of his in case a challenge thereto should arise later.

11.2. For the purpose of Clauses 11.3 to 11.5 inclusive a dispute is deemed to arise when one party serves on the other a notice in writing (herein called the Notice of Dispute) stating the nature of the dispute. Provided that no Notice of Dispute may be served unless the party wishing to do so has first taken any step or invoked any procedure available elsewhere in the Contract in connection with the subject matter of such dispute and the other party or the Engineer as the case may be has:-

(a) taken such steps as may be required

or

(b) been allowed a reasonable time to take any such action.

It is often extremely difficult to ascertain precisely when a particular dispute first arose. Thus the Contractor might apply to the Engineer for permission to do work in a certain way, or for additional cost or an extension of time and be refused, either wholly or in part, yet no dispute will arise unless the Contractor is unwilling to accept such refusal. Similarly, the Engineer may agree to something to which the Employer objects, but that objection may not be realized or perceived until a later stage. Again, in the 'to and fro' of negotiation, proposals from one side are often rejected for the time being by the other, but the result is more negotiation, not a dispute.

For this reason Clause 11.2 lays down a firm definition of when a dispute is to be deemed to come into existence, namely the service by one party upon

---

1. A tort is a civil wrong independent of contract. A person's duty to 'the world at large' can, however, also be owed to a person with whom the first person has a contract. A given event can therefore give rise to liability both for breach of contract and for breach of some general duty in tort. (There may well in some cases be criminal liability as well.)

the other of a written Notice of Dispute. Thus the rule under Clause 11 is very simple — no notice, no dispute. Moreover, the notice may be served at any time, unlike the position under some other sets of standard conditions when it may only be served after the Works have been completed.

However, a dissatisfied party should be slow to issue any notice unless and until all hope of a settlement by negotiation is at an end or where, as may sometimes be the case, resolution of the dispute is impeding the further progress of the Works. Needless to say, a flood of ill-considered notices during the course of the Works is hardly likely to make for good (and therefore efficient and profitable) contractual relationships.

The notice must not only be in writing but must also state the nature of the dispute. It is of greatest importance that such statement should describe concisely and clearly what the dispute really is about. An accurate description will improve the chances of a reasonable solution by negotiation, while an inaccurate or incomplete description may well hamper future procedures since any eventual arbitrator's jurisdiction (authority to decide) will be largely confined to the dispute as stated in the notice.

While notices may be issued at any time as disputes arise, the issuing party cannot do so unless and until all other means of resolving the situation giving rise to the dispute have been tried and have failed. The words of the clause are clear and unequivocal and need no further elaboration here.

Failure of either party to act on a Notice of Dispute

If within 28 days of the issue of a Notice of Dispute there is no reference of the matter either to conciliation or to arbitration, the Notice of Dispute becomes void. The situation thereafter is summarized in Note 14(3).

> NG14(3) It should be noted that if within the prescribed period of 28 days after service of a Notice of Dispute neither party has made a request in writing for the dispute to be referred to a conciliator nor has served a written Notice to Refer requiring the dispute to be referred to arbitration then the Notice of Dispute becomes void. It is then open to either party to continue the dispute by serving a fresh Notice of Dispute unless the first Notice of Dispute was served within the 28 days allowed under Clause 7.7 (Final certificate) in which case the final certificate becomes final and binding and no further dispute in respect of the Contract is possible.

The Notice of Dispute is of particular relevance in connection with the final certificate under Clause 7.7. This certificate is to be 'conclusive evidence as to the sum due to the Contractor under or arising out of the Contract . . . unless either party has within 28 days after the issue of the final certificate given notice under Clause 11.2'. The importance of this provision is that after the 28 days both parties will be time-barred from disputing the final certificate which thereafter becomes final and binding upon both.

Conciliation

> 11.3. In relation to any dispute notified under Clause 11.2 and in respect of which no Notice to Refer under Clause 11.5 has been served either party may within 28 days of the service of the Notice of Dispute give notice in writing requiring the dispute to be considered under the Institution of Civil Engineers' Conciliation Procedure (1988) or any amendment or modification thereof being in force at the date of such notice

and the dispute shall thereafter be referred and considered in accordance with the said procedure.

As mentioned above, conciliation is the minor works form's replacement for the safety valve condition precedent in ICE 5 of the 'Engineer's Clause 66 Decision'. Once matters have reached the stage that, first, all attempts to resolve the situation have failed and, second, that a Notice of Dispute under Clause 11.2 has been issued, experience shows that some means of postponing formal arbitration to allow a cooling-off period is very often instrumental in resolving the difficulty or, at least, allowing the parties to consider their position in a wider context so that the chance of a commercial (rather than a purely legal) solution is improved.

Conciliation is optional, and either party can elect to proceed directly to arbitration at any time. But, given that a Notice of Dispute has validly been issued, and provided that neither side has then issued a Notice to Refer (to arbitration; see below), the parties then have 28 days from the date of issue of the Notice of Dispute in which to refer the matter to conciliation. Other than that it must be in writing, the notice to refer to conciliation can take any convenient form, save only that it must, of course, accurately identify the dispute(s) to be so referred.

The prior issue of a Notice to Refer under Clause 11.5 will prevent that dispute from being referred to conciliation at all. In practice, therefore, once a Notice of Dispute has been issued, it will be the first notice to be served (for conciliation or for arbitration) which will prevail, and a party with a preference for conciliation would be wise to issue the notice to refer to conciliation at the same time as the relevant Notice of Dispute is issued.

NG14(2) It is normally expected that the party serving a Notice of Dispute under Clause 11.2 will at the same time serve notice in writing either under Clause 11.3 (Conciliation) or Clause 11.5 (Arbitration).

It is clearly preferable, however, that the parties should agree on the course to be taken.

Conciliation procedure

Conciliation is not defined in the minor works contract. The procedure to be followed is, however, laid down as that contained in the Institution of Civil Engineers' Conciliation Procedure. This procedure is set out in full in Appendix C of this book and is discussed fully in Chapter 19. As stated in Note 14 in the Notes for guidance, the conciliation is complete when the conciliator has delivered his recommendations and, if any, his opinion to the parties.

NG14(1) The option provided by the Conciliation Procedure under Clause 11.3 is intended to provide a means whereby disputes can be settled with a minimum of delay by obtaining an independent recommendation as to how the matter in dispute should be settled. The conciliation is complete when the conciliator has delivered his recommendations and, if any, his opinion to the parties.

Hopefully, that will end the dispute either by the parties' acceptance of the conciliator's recommendations or by some other settlement by negotiation, using the conciliator's recommendations or opinion as a starting point. If not, the provisions of Clause 11.4 or 11.5 will apply.

Arbitration

Whether or not conciliation has been attempted, the parties may always resort to arbitration. In either event, the minor works form provides that reference to arbitration must be made within 28 days, which period runs either from the date of the conciliator's recommendation or (if conciliation is not used) from the service of the original Notice of Dispute. The consequences of failing to act within the 28 days do, however, differ.

> 11.4. Where a dispute has been referred to a conciliator under the provisions of Clause 11.3 either party may within 28 days of the receipt of the conciliator's recommendation refer the dispute to the arbitration of a person to be agreed upon by the parties by serving on the other party a written Notice to Refer. Where a written Notice to Refer is not served within the said period of 28 days the recommendation of a conciliator shall be deemed to have been accepted in settlement of the dispute.

Failure to serve the Notice to Refer within the prescribed 28 days will therefore render the terms of the conciliator's recommendation final and binding on both parties. In this it is analogous to the formal decision of the Engineer under ICE 5, Clause 66(1) which similarly becomes irrevocable once the stipulated time for challenging has been allowed to expire.

> 11.5. Where a dispute has not been referred to a conciliator under the provisions of Clause 11.3 then either party may within 28 days of service of the Notice of Dispute under Clause 11.2 refer the dispute to the arbitration of a person to be agreed upon by the parties by serving on the other party a written Notice to Refer. Where a Notice to Refer is not served within the said period of 28 days the Notice of Dispute shall be deemed to have been withdrawn.

Where the parties prefer to proceed directly to arbitration, failure to do so by issuing a Notice to Refer within 28 days of the original Notice of Dispute merely results in the deemed withdrawal of the latter notice. It will then normally be open to either party to serve a fresh Notice of Dispute, thereby obtaining a further 28 day period in which to proceed to arbitration, or indeed conciliation. The issuing party may well wish to serve the Notice to Refer with the fresh Notice of Dispute.

The only exception is where the original Notice of Dispute was against the final certificate which, by Clause 7.7, becomes 'conclusive evidence as to the sum due to the Contractor' if it is not challenged within 28 days. Thus, if the original Notice of Dispute is allowed to lapse, it may then be too late to issue a valid replacement notice.

Appointment of arbitrator

> 11.6. If the parties fail to appoint an arbitrator within 28 days of either party serving on the other party a written Notice to Concur in the appointment of an arbitrator the dispute shall be referred to a person to be appointed on the application of either party by the President (or if he is unable by any Vice President) for the time being of the Institution of Civil Engineers.

Commentary

1. By agreement

As with a conciliator, the normal method of appointment should be for the parties to agree on whom they want as arbitrator and then invite that person to accept the appointment. For this purpose the parties will often know of an acceptable person, or if they do not they can consult the ICE List of Arbitrators.

Unfortunately, the usual practice now seems to be that the party issuing the Notice to Refer will suggest a number of persons (usually three) whom they consider suitable which the other party promptly rejects out of hand, putting forward his own three which the original party then rejects in his turn. This nonsensical situation is probably the result of corresponding at arm's length (which almost always happens when lawyers are involved) and can often be avoided by a simple face-to-face meeting of principals of the parties either arranged for the express purpose of choosing an arbitrator or (more conveniently) as part of one called for other purposes, such as a routine progress meeting.

Only if such a face-to-face meeting cannot result in agreement should the alternative of issuing a Notice to Concur be used, since the main reason for its inclusion in the minor works form is to meet the situation where one party simply refuses to discuss any appointment at all (usually as a delaying tactic).

There is, of course, no reason why a Notice to Concur should not be issued with the Notice to Refer, since a face-to-face meeting can (and should) still take place. The advantage is that if agreement is not reached within 28 days of the concurrent issue of notices, application can then be made to the President of the Institution of Civil Engineers for an appointment without further loss of time. Indeed, there is no reason why both notices cannot form part of a single document. However, it is strongly recommended that only one name be entered in the Notice to Concur. The reason for this is the practical one that, if three names are to be inserted and it seems likely that the other party will reject the first party's proposed names out of hand, it may seem logical to put in the names of three 'low-grade' arbitrators so that the person the proposing party really wants can still be available for appointment by the President.

Should the other party unexpectedly agree to one of these names, however, the proposing party will be stuck with an arbitrator that he considers to be less than suitable. On the other hand, if the proposing party puts forward his three preferred arbitrators, all three might be ruled out. Indeed, the six 'best' arbitrators (from an objective viewpoint) can often be ruled out by this procedure and, in any single area of specialization, there may remain no really suitable arbitrators at all.

It must follow that only one name should be entered into any Notice to Concur. It is sometimes argued (often by solicitors) that proper procedure or custom makes the entry of three names obligatory. *There is no such rule*, and suggestions to the contrary should be ignored.

2. By the President

Should the Notice to Concur fail to resolve the difficulty, either party (not just the one who issued the notice) can apply to the President of the Institution of Civil Engineers for him to make an appointment. Such application cannot normally be made until 28 days after the notice was received by the party to whom it was addressed, but the parties are at liberty to agree to apply at any time, in which case the 28 day limit will not be relevant and there may not even be a need to issue a Notice to Concur.

Indeed, the parties can, if they do not wish to embark on conciliation, agree in advance that application to the President of the Institution of Civil Engineers can be made at the same time as the original Notice of Dispute, in order to save time, but such cases will be rare. It should be noted, however, that such an application cannot be made before the Notice of Dispute is served, since by Clause 11.2 a dispute cannot be said to arise before then and the President will not normally make an appointment in respect of a dispute which legally does not yet exist.

The procedure for making applications to the President is laid down in Part A of the ICE Arbitration Procedure (1983) and the forms appropriate for so doing are obtainable from the Arbitration Office, The Institution of Civil Engineers, 1–7 Great George Street, London SW1P 3AA. A copy of this is reproduced in Appendix E.

The President will normally appoint a person whose name appears in the current ICE List of Arbitrators but he is at liberty to go outside that list should the circumstances of a particular application make it advisable to do so. In this connection it is often asserted that the President cannot appoint a person whom one of the parties has already rejected, but there is no legal authority for this proposition. The President will naturally try to find someone who has not been rejected by one of the parties but, if the result would be the appointment of a really unsuitable arbitrator, the better view is that the President can appoint whomsoever he likes, regardless of any prior rejection, provided that there is no other objection on the grounds of bias or conflict of interest.

There is, however, one specific limitation on the President's power of choice. Where a reference to arbitration follows an unsuccessful attempt at conciliation, Rule 14 of the ICE Conciliation Procedure (1988) precludes the President from appointing that conciliator as arbitrator in respect of the same disputes. On the other hand, the parties can, if they so wish, ask the conciliator to accept appointment as arbitrator. However, such an appointment could present a number of difficulties which are probably better avoided. These difficulties are discussed in the legal comment at the end of this chapter.

Arbitration procedure

All arbitrations under the minor works form must be conducted in accordance with the ICE Arbitration Procedure (1983) unless the parties otherwise agree. Moreover, unlike the situation under ICE 5, Clause 66, the Short Procedure (Part F of the ICE Arbitration Procedure (1983))[1] is not optional but must be followed unless the parties agree otherwise in writing.

1. The Short Procedure is set out in Appendix F.

11.7. Any such reference to arbitration shall be conducted in accordance with the Institution of Civil Engineers' Arbitration Procedure (1983) or any amendment or modification thereof being in force at the time of the appointment of the arbitrator and unless otherwise agreed in writing shall follow the rules for the Short Procedure in Part F thereof. Such arbitrator shall have full power to open up review and revise any decision instruction direction certificate or valuation of the Engineer.

It will be noted from the last sentence of this clause that the familiar power of the arbitrator under ICE 5 to 'open up review and revise any decision instruction or direction certificate or valuation of the Engineer' applies also to the resolution of disputes under the minor works form.

This Short Procedure is discussed at length in Chapter 9 of *The Institution of Civil Engineers' Arbitration Practice*[1] The following short summary indicates the broad outline which minor works arbitrations will usually follow. The rules here quoted are those of the ICE Arbitration Procedure (1983).

Statements of case

Rule 20.2 Each party shall set out his case in the form of a file containing

(*a*) a statement as to the orders or awards he seeks

(*b*) a statement of his reasons for being entitled to such orders or awards

and

(*c*) copies of any documents on which he relies (including statements) identifying the origin and the date of each document

and shall deliver copies of the said file to the other party and to the Arbitrator in such a manner and within such time as the Arbitrator may direct.

Formal pleadings are not required, and the key evidence must be included in the file. The statements to be included under sub-paragraph (*c*) are witness statements. The orders or awards referred to under sub-paragraphs (*a*) and (*b*) will usually be the sums of money which the party concerned hopes to receive, but can also include declarations as to liability or even simple findings of fact. The statement of reasons under sub-paragraph (*b*) is equivalent to the argument and submissions which an advocate would deliver in court or in a full arbitration.

The intention behind Rule 20.2 is that each party's complete case (claims, submissions and evidence) shall be prepared in parallel and that the resulting files shall be sent to the arbitrator in duplicate and at the same time. The arbitrator (having first examined the files for procedural errors, gaps or inconsistencies) should then send on to the other party one copy of each party's file, retaining the other copy for his own use and thereby ensuring that everyone has the same set of papers.

However, the respondent party will often ask to see the claimant party's file before he prepares his own, in which case this part of the procedure will take a little longer to complete. In any event, and whether the files are delivered at the same time or one after the other, it will usually be appropriate for the arbitrator to offer each party a brief opportunity to comment in writing on the other's file.

---

1. By Hawker, Uff and Timms (1986), London: Thomas Telford.

Site visits

Rule 20.3 After reading the parties' cases the Arbitrator may view the site or the Works and may require either or both parties to submit further documents or information in writing.

If the arbitrator does view the Site he should do so either on his own or in the company of a representative from each party. What he must never do is to make his site visit with one party present and the other absent. A site visit can often be very useful, since it allows the arbitrator to acquaint himself with the circumstances of the dispute and enables him to bring his own professional expertise into play. Relevant information gleaned on such a visit is real evidence and need not be spelt out expressly in the parties' files, provided always that the arbitrator lets each party know of such facts as he intends to take into account.

Further information

Having read the parties' files and (perhaps) having visited the Site, the arbitrator may ask the parties to submit further evidence, comment or submissions in writing. In responding to such a request, the parties' replies must be restricted to those matters about which the arbitrator is enquiring; it is not an opportunity for raising new claims, producing fresh evidence or submitting further arguments in general. The parties' replies will be added to their respective files, ready for the meeting under Rule 20.4.

The meeting

Rule 20.4 Within one calendar month of completing the foregoing steps the Arbitrator shall fix a day when he shall meet the parties for the purpose of

(*a*) receiving any oral submissions which either party may wish to make and/or

(*b*) the Arbitrator's putting questions to the parties their representatives or witnesses. For this purpose the Arbitrator shall give notice of any particular person he wishes to question but no person shall be bound to appear before him.

This meeting is not a hearing. Its purpose is to let the arbitrator test the evidence and take the parties' arguments further. Thus the parties are neither entitled to lead fresh evidence, nor may they examine the witnesses themselves. They do have the right, however, to make further oral submissions if they so wish.

Accordingly, the arbitrator will tell the parties in advance which, if any, of the witnesses whose statements are already on file he wishes to attend the meeting. If they appear (and the arbitrator has no power to compel them to come) the arbitrator will himself examine them, normally on oath, and the parties (or their advocates) may not intervene unless the arbitrator gives them leave. Their right to make oral submissions should therefore normally be exercised after the arbitrator has finished with all the witnesses.

The foregoing is what might be called the orthodox procedure at the meeting, and the one which will almost certainly be observed if lawyers are in attendance. However, the meeting can often with advantage be allowed to develop into a round-table discussion, with the witnesses arguing and the arbitrator leading the discussion and identifying points as they emerge.

Commentary

Whether or not such discussion should be enlarged to include the parties or their advocates is a matter that the arbitrator will have to decide as he goes along. But in any event, where such informal discussion does take place, the arbitrator should always summarize the results before closing the meeting and invite the parties or their advocates to contribute short comments on his summary.

The award

Rule 20.5 Within one calendar month following the conclusion of the meeting under Rule 20.4 or such further period as the Arbitrator may reasonably require the Arbitrator shall make and publish his award.

Unless the issues before him are unusually complex, one calendar month should normally be adequate for the preparation and publication of an award on a minor works dispute. There is no legal limit to the time the arbitrator may take to make his award but, as a rough guide, anything over three months will normally be considered excessive by the courts.

Costs

Rule 21.1 Unless the parties otherwise agree the Arbitrator shall have no power to award costs to either party and the Arbitrator's own fees and charges shall be paid in equal shares by the parties. Where one party has agreed to the Arbitrator's fees the other party by agreeing to this Short Procedure shall be deemed to have agreed likewise to the Arbitrator's fees.

The principle is that each party shall pay his own costs in the arbitration and one half of the arbitrator's fees and expenses, regardless of the result. It follows that if a party wishes to engage lawyers or expensive expert witnesses he must do so at his own expense and cannot pass on the costs so incurred to the other party.

This is a sensible provision, since it should normally encourage the parties to present their cases concisely and with minimum complications, thereby facilitating a prompt and sensible solution of the matters in dispute. On the other hand, should one party deliberately deploy claims experts or indulge in other obstructive tactics, the other party may well feel compelled to protect himself in like manner, despite having no prospect of being able to recover the expenditure so incurred at the end of the day. In an extreme case a party who is subjected to this kind of harassment (which the arbitrator may be hard put to control) might be driven into insolvency before he can have his dispute resolved.

There is in any event another problem, namely that Section 18.3 of the Arbitration Act 1950 provides that any agreement that each party shall pay their own costs is void unless the agreement is entered into *after* the dispute has arisen. The wise arbitrator will therefore seek to obtain the parties' written confirmation of Rule 21.1 before proceeding with the reference.

Other procedures

However, Rule 21 of the Short Procedure also allows either party to give written notice to the other requiring that the arbitration revert to full oral procedure (Rule 21.2), but Rule 21.3 stipulates that the party making the request has to pay the costs of the arbitrator, and reasonable compensation to the other party for his costs, that have already been incurred.

> Rule 21.2 Either party may at any time before the Arbitrator has made and published his Award under this Short Procedure require by written notice served on the Arbitrator and the other party that the arbitration shall cease to be conducted in accordance with the Short Procedure. Save only for Rule 21.3 the Short Procedure shall thereupon no longer apply or bind the parties but any evidence already laid before the party shall be admissible in further proceedings as if it had been submitted as part of those proceedings and without further proof.
>
> Rule 21.3 The party giving notice under Rule 21.2 shall thereupon in any event become liable to pay
>
> (*a*) the whole of the Arbitrator's fees and charges incurred up to the date of such notice
>
> and
>
> (*b*) a sum to be assessed by the Arbitrator as reasonable compensation for the costs (including any legal costs) incurred by the other party up to the date of such notice.
>
> Payment in full of such charges shall be a condition precedent to that party's proceeding further in the arbitration unless the Arbitrator otherwise directs. Provided that non-payment of the said charges shall not prevent the other party from proceeding in the Arbitration.

It is arguable that, since the minor works form provides only for arbitration under the Short Procedure, these two rules are incompatible with the Contract and are thus of no effect. It would therefore follow on this argument that a party cannot unilaterally terminate the Short Procedure. On the other hand, the parties remain free to agree on another procedure if they so wish. Unfortunately, the minor works form is silent on this aspect, and it is by no means certain what line the court might choose to take on appeal.

For disputes which are primarily technical in nature, however, there may well be merit in considering the use of the Special Procedure for Experts under Part G of the ICE Arbitration Procedure (1983) as an alternative to the Short Procedure, which, nevertheless, it largely resembles. Those wishing to adopt this alternative should refer to *The Institution of Civil Engineers' Arbitration Practice.*[1]

Comparison with ICE 5

Clause 66 of ICE 5 contains a condition precedent to arbitration in that any dispute or difference must first be referred back to the Engineer for a formal decision under that clause (whether or not he has already decided the same matter under some other Clause of the Contract). If the Works are still in progress the Engineer then has one month within which to give his formal decision, while if a completion certificate has already been issued he has three months in which to do so. Arbitration must then be commenced, if at all, within three months from the date of the Engineer's formal decision

1. By Hawker, Uff and Timms (1986), pp. 87–90, London: Thomas Telford.

or, if the Engineer fails to give his decision within the time allotted, within three months after the expiry of that period.

This condition precedent was considered to be too onerous for the average minor works project, the more so that up to six months could elapse before the threat of arbitration is lifted. On the other hand, the condition precedent was seen to be useful in that it provided a cooling-off period and an opportunity for re-appraisal before the parties became locked into arbitration.

The solution adopted was therefore to omit the condition precedent but to introduce in its place an optional and informal conciliation procedure. Unlike the Engineer making a Clause 66 formal decision, the conciliator is entitled to depart from the strict terms of the Contract but, like the Engineer's decision, the conciliator's recommendation becomes final and binding if not challenged within the period prescribed in the Contract.

With regard to arbitration itself, both ICE 5 and the minor works form require the reference to be conducted under the ICE Arbitration Procedure (1983), but under the minor works form this is further restricted to the use of the Short Procedure in Part F.

Comparison with ICE 6

The draughtsmen of ICE 6 have retained virtually unaltered (albeit with sub-clauses re-numbered) both the condition precedent for a formal Engineer's decision and the ICE 5 arbitration sub-clauses as they stand. However, the minor works requirement for an initial Notice of Dispute has been adopted for ICE 6 and an optional conciliation facility inserted between the condition precedent and full arbitration.

Legal comment

Under English law, the courts have a discretionary power[1] to stay litigation on any matter covered by a valid arbitration agreement. An arbitration agreement does not prevent one of the parties from starting an action in the courts but, under section 4(1) of the Arbitration Act 1950, the Defendant can seek a stay provided that application is made before any procedural step beyond mere acknowledgement of service of the writ or originating motion or summons has been taken. Thus a party wishing to obtain a stay must not deliver a defence to the action nor agree to any order for directions before applying for the action to be stayed.

The courts will not normally refuse a stay. However, a stay may be refused if

(i) the matters at issue are almost wholly matters of law which can conveniently be dealt with by the court (or which are likely to be the subject of an eventual appeal from the arbitrator's award; or
(ii) there are several parties to the dispute, not all of whom are bound by the arbitration agreement; or

1. There is no such discretion for 'non-domestic' disputes — i.e. in general, those having an element outside England and Wales — since under the 1975 Act the court must grant a stay.

(iii) there are allegations of serious impropriety or fraud, in which case those so accused have a common law right to trial and (hopefully) vindication in public; or
(iv) no dispute really exists.

Arbitration law is at present under review, and arbitrators may well be given wider powers than they now enjoy, with the courts taking a purely supportive role rather than a more 'interventionist' stance as in the past. This is not, however, the place to discuss such developments.

Methods of so-called alternative disputes resolution (ADR) have recently achieved some prominence, although their form and even nomenclature is still often unclear. However, the conciliation provisions of the minor works form are clearly among these methods.

Finally, should the parties decide that they wish their conciliator to become their arbitrator (and assuming that he agrees), it should be emphasized that, to ensure that future procedural and/or jurisdictional problems will not arise, there should be an express agreement to that effect in writing, and that this agreement should stipulate clearly to what extent the arbitrator may have regard to evidence which was put to him as conciliator. Nor should it be forgotten that the Notice to Refer under Clause 11.4 must still be served within the time limits prescribed, and that the award of the arbitrator will be final and binding in a manner that any earlier recommendation of the conciliator was not.

Chapter 15

JURISDICTIONS OTHER THAN ENGLAND AND WALES

Introduction

Throughout this book it has been assumed that the minor works form is to be used in connection with projects in England and Wales and that the requirements of English law therefore apply. There are, however, within the United Kingdom a number of other legal jurisdictions, that is legal systems for the administration of justice. The other main ones are Scotland and Northern Ireland, but there are different systems in the Isle of Man and the Channel Islands. Each jurisdiction has its own legal requirements.

There is. of course, nothing to stop the minor works form from being used in foreign jurisdictions. The widely-used FIDIC Conditions are largely based upon the ICE Conditions and are used throughout the world. Many countries, such as those within the Commonwealth and the United States, have legal systems based upon the English common law.

Before using the minor works form for projects outside the United Kingdom and the Republic of Ireland it would be prudent to check that its provisions — in particular those concerned with dispute settlement — are fully compatible with the local law. Alternatively, an express provision could be written in stating that the proper law of the contract shall be the law of England and Wales, but even then it would be wise to take legal advice before proceeding.

Scotland

Application in Scotland is covered by Clause 12.

12.1. If the Works are situated in Scotland the Contract shall in all respects be construed as a Scottish contract and shall be interpreted in accordance with Scots Law and the following Clause shall apply.

Clause 12.1 merely makes the obvious point that if the Works are situated in Scotland Scots law will apply. However, it is perfectly feasible to stipulate (by means of an express special condition) that English law shall apply even for works in Scotland. This is hardly ever done since it could lead to difficulty.

In practice Scots law makes very little difference to the manner in which works under the minor works form are actually carried out, and the running of the Contract.

12.2 In Clause 11 hereof the word 'arbiter' shall be substituted for the word 'arbitrator' and 'the Institution of Civil Engineers' Arbitration Procedure (Scotland) (1983)' shall be substituted for 'the Institution of Civil Engineers' Arbitration Procedure (1983)'.

This clause is necessary to adapt Clause 11 (Disputes) to Scottish procedures, and then only changes the nomenclature so the the ICE Arbitration Procedure designed for use in Scotland is made applicable.

A full commentary on this procedure can be found in Chapters 11 and 12 of *The Institution of Civil Engineers' Arbitration Practice.*[1]

Northern Ireland

The law of Northern Ireland is largely the same as English law, and a minor works project in that country will be conducted in the same manner and upon the same legal basis as one in England and Wales. The only differences of significance are in arbitration, and a full explanation of the necessary procedures is set out in Chapter 13 of *The Institution of Civil Engineers' Arbitration Practice.*

Southern Ireland

Before 1921 the law of 'Southern' Ireland was the same as for Northern Ireland. Since the establishment of the Republic, however, the two systems of law have diverged somewhat and, in particular, the Republic of Ireland now has a written constitution which will sometimes affect the legal construction to be put upon provisions which otherwise seem identical.

The Institution of Engineers of Ireland (IEI) has also produced its own IEI Conditions of Contract and an IEI Arbitration Procedure. Whilst these documents differ in some respects from their ICE equivalents the minor works form can still be considered a simplified version of the IEI Conditions.

Accordingly, the comments in respect of Northern Ireland apply also to projects in the Republic of Ireland, with the proviso that there remain some slight differences between the laws of the two countries. Further comment can be found in Chapter 13 of *The Institution of Civil Engineers' Arbitration Practice.*

Other jurisdictions

The minor works form can safely be used in other parts of the United Kingdom, such as the Channel Islands and the Isle of Man. It can probably also be used with success in foreign jurisdictions where the local law is derived from the English common law, such as the Commonwealth or the United States of America. It would, however, be prudent to take legal advice on the effects (if any) of the local law, in particular with respect to the procedures for settling disputes. The reasons for this are explained in Chapter 15 of *The Institution of Civil Engineers' Arbitration Practice.*

In all jurisdictions the ICE Conciliation Procedure 1988 may be used without alteration as the resulting recommendations are not enforceable at law.

---

1. By Hawker, Uff and Timms (1986), London: Thomas Telford.

Part 3

APPLICATION

Chapter 16

OTHER STANDARD DOCUMENTS

Introduction

In contracts using the minor works form it will often be necessary to use other standard forms in association with it, particularly a standard method of measurement when compiling a Bill of Quantities.

Two such forms are particularly useful: the Civil Engineering Standard Method of Measurement (CESMM 2) and the Federation of Civil Engineering Contractors' daywork schedules. Both these documents have been prepared for use with ICE 5 and need slight amendment in order to be compatible with the minor works form.

Standard methods of measurement — CESMM

Bills of Quantities for use with the minor works form may be compiled in accordance with The Civil Engineering Standard Method of Measurement (second edition), commonly known as CESMM 2, provided that a few amendments are made to it. These amendments are necessary because the CESMM 2 was written to be used in conjunction with ICE 5, and the minor works form differs in three distinct ways: the definitions vary in detail, the clause numbers in the two documents for the same subject are different, and nominated sub-contractors are not permitted in the minor works form.

CESMM 2 provides that any deviations from it must be expressly stated in the preamble to the Bills of Quantities, and a sample preamble to make CESMM 2 compatible with the minor works form is provided in Appendix E.

The rules for measurement in CESMM 2 do not depend upon the conditions of contract for their interpretation, so if this sample preamble is incorporated into the Bill of Quantities, that will be the only difference between a Bill for use with the minor works form and one for use with ICE 5 or 6. This is an advantage because, should it be decided at a late stage in the pre-tender programme to change from using ICE 5 to the minor works form, it will generally only mean that it is the preamble which requires amendment and not the sections of the Bill.

Amendments to work classifications

1. Method-related charges

For minor works projects it may not be considered desirable for the Contractor to be permitted to enter method-related charges into the Bill of Quantities, or to provided separately for general items such as insurance or offices for the Resident Engineer. Both these items are incorporated in a Bill of Quantities for two main reasons each of which have benefits for both the Employer and the Contractor. They do, however, increase the administrative work involved in measurement and certification, and this may well cancel out the benefits for minor works contracts.

Method-related charges are explained in Section 7 of CESMM 2. Basically they are items that the Contractor enters into the Bill of Quantities at the time of tender for expenditure which is not dependent upon the quantities of permanent works carried out. They are either fixed sums (such as the provision of the Contractor's workshops, site compounds or access roads), or time-related amounts (such as the cost of running of the Contractor's office, the supervision of the Works, or the provision of water, electricity and telephones).

The alternative to method-related charges is for the Contractor to spread these sums amongst the other items of work in the Bill of Quantities. This will have one of two effects. Firstly, in the event of variations the price paid by the Employer will be determined by the bill rates. If these contain a spread of a fixed lump sum then the amount paid by the Employer will be higher than if the lump sum is billed separately. Secondly, the total expenditure will be reimbursed to the Contractor not when the expenditure is incurred but when the item of work upon which it is spread (and to which its occurrence is not related) is actually executed. This will usually result in the Contractor being paid much later than the time at which the expenditure is incurred. The Contractor will have to finance the resulting payment shortfall, and the cost of such financing will be included in the tender price at the interest rate at which the Contractor borrows the money even though the Employer can usually borrow more cheaply than the Contractor.

When the temporary works designed by the Contractor are small then the administrative effort involved in assessing payment for method-related charges may outweigh the benefits to be gained from their use.

2. *General items*

General items (Class A in CESMM 2) are included for the same reasons. The general items cover the cost of the provision of insurance, and the cost of the requirements of the Engineer (such as offices, assistance, and tests). There is no reason why items for these costs, which are likely to be small on a minor works contract, should not also be omitted.

The Employer benefits from including these items by

(i) more realistic rates for variations, and
(ii) reduced financial cost.

The Contractor benefits from improved cashflow.

For a minor works contract the amount of variations should be small, and so should the cost of the Engineer's requirements. Consequently the omission of general items will not greatly affect either party. However, if the general items are omitted, then any temporary works designed by the Engineer, and any other specified requirements must be billed under the other classes in the work classification.

FCEC dayworks schedules

The full title of the FCEC dayworks schedule is 'The Federation of Civil Engineering Contractors' Schedules of Dayworks carried out incidental to Contract Works'. The dayworks schedules are intended to be applicable for dayworks ordered to be carried out 'incidental' to the Contract Works. They are not intended to be applicable to dayworks ordered to be carried out after the Contract Works have been substantially completed, or to a contract to be carried out wholly on a daywork basis.

The circumstances of such works vary so widely that the rates applicable call for special consideration and agreement between Contractor and employing authority.

Use of the schedules

The schedules consist of four sections.

1. Labour

The general principle is that the Contractor is paid the 'amount of wages' paid to workmen, together with a percentage addition. The addition is either the one that is given in the schedule, or alternatively the Contractor is given the opportunity, in the tender documents, to offer the percentage addition that he requires.

The amount of wages is defined as wages, and

(i) actual bonus paid,
(ii) daily travelling allowances (fare and/or time),
(iii) tool allowance and
(iv) all prescribed payments, including those in respect of time lost due to inclement weather paid to workmen at plain time rates and/or at overtime rates.

The percentage addition is to cover all statutory charges and other charges, including

(i) National Insurance and surcharge;
(ii) normal contract works, third party and employers' liability insurances;
(iii) annual and public holidays with pay and benefit scheme;
(iv) non-contributory sick-pay scheme;
(v) industrial training levy;
(vi) redundancy payments contribution;
(vii) Contract of Employment Act costs;
(viii) site supervision and staff, including foremen and walking gangers, but the time of gangers or chargehands working with their gangs is to be paid for as workmen;
(ix) small tools such as picks, shovels, barrows, trowels, handsaws, buckets, trestles, hammers, chisels and all items of a like nature;
(x) protective clothing; and
(xi) head office charges and profit.

Labour-only sub-contractors are paid at full invoice, plus an on-cost percentage (currently 88%). Subsistence, lodging or period travel allowances, if paid, are charged at cost plus a percentage addition (currently 12%). The general principle is that the cost of site overheads is an allowance on the cost of wages and labour-only sub-contractors' accounts.

When assessing the time for dayworks, it is always necessary to remember to take into account mobilization and demobilization while the labour is moving from measured work to dayworks, and from dayworks back to measured work.

2. *Materials*

Materials are paid for at cost delivered to Site. This means the invoiced cost without the deduction of any discount for cash not exceeding $2\frac{1}{2}$%. To this figure is added a percentage on-cost (currently $12\frac{1}{2}$%) to cover head office charges and profit. Where materials have to be unloaded, any labour or plant involved is paid for at the daywork rates in the schedule.

3. *Plant*

The major part of the FCEC dayworks schedule is devoted to plant rates (26 pages out of 39). Most types of plant encountered in civil engineering are listed with an appropriate hire rate, normally by the hour, but occasionally by the day or week. The hire rates are intended to cover all plant on the Site, whether in the ownership of the Contractor or hired in. Where plant is especially hired in for dayworks, then there is an additional percentage on-cost.

4. *Supplementary charges*

There are a number of items, as defined in this schedule, for which specific percentages are applied:

(i) the cost of free transport provided by the Contractors for workmen to and from the Site (currently $12\frac{1}{2}$%);
(ii) sub-contractors' accounts (currently $12\frac{1}{2}$%);
(iii) the cost of internal transport on the Site (currently the appropriate daywork rates);
(iv) the cost of operating welfare facilities (currently $12\frac{1}{2}$%);
(v) the cost of any additional insurance premiums for abnormal contract work or special site conditions (currently $12\frac{1}{2}$%); and
(vi) the cost of watching and lighting, specially necessitated by dayworks (currently nett cost).

Where VAT is applicable, it is charged in addition to the rates in the schedule in the amount payable to HM Customs and Excise.

Other uses

The circumstances where it is desirable that the Works should be carried out on a daywork basis vary so widely that the rates applicable call for special consideration according to the circumstances.

1. Projects wholly on a daywork basis
Where the whole of the Contract is to be carried out on dayworks the Contractor should always be permitted to apply his own percentages to the costs of labour, materials and plant, as defined in the schedules. The method and layout of applying the percentages is best done in accordance with paragraph 5.6 of CESMM 2 (e.g. method 2 in Appendix E1, page 268). The reason for this is that the daywork rates must support the cost of the Contractor's entire site organization, whereas the FCEC dayworks schedules are normally intended to cover only the site administration necessary to administer dayworks carried out incidental to other Contract Works.

2. Dayworks after substantial completion
The Contractor should also be given an opportunity to provide his own on-cost percentages, where major items of dayworks occur after substantial completion or at times when no measured Contract Works are being carried out concurrently, and where the dayworks themselves do not form a substantial part of the Contract. Again, the dayworks rates must support the Contractor's entire site administration costs for the period when dayworks are the only work in hand.

General

Since the FCEC dayworks schedules were first produced there has been a significant change in the working practice of the construction industry in that most operatives are now being employed by sub-contractors rather than being directly employed by the Contractor. The principle of recovering the site administrative costs by means of an on-cost percentage to the cost of wages is therefore somewhat anachronistic. On the other hand, to reflect current market conditions, Contractors may wish to spread their administrative costs more evenly over all their expenditure. If so, the percentages to be applied are likely to show an increase on all items except labour.

Suggested layouts for the inclusion of daywork charges into the Bills of Quantities are given in Appendix E1. Copies of the FCEC dayworks schedule can be obtained from the FCEC or the ICE.

Chapter 17

SUB-CONTRACT DOCUMENTATION

Introduction

There are no published sub-contract forms designed specifically for use with the minor works form. The simplest form of sub-contract is an order describing the work to be carried out and agreeing a payment. The major terms of such a contract will often be in a form derived from the Supply of Goods and Services Act 1982, which in turn will import a number of implied terms into the Contract. However, such a contract is not to be recommended because it fails to address a number of essential points such as the power of the Contractor to give instructions, the insurances the sub-contractor is required to take out, damages and, from the sub-contractor's point of view, the relief given if unforeseen physical conditions are encountered.

The proper and sensible approach is for the Contractor to enter into a 'back to back' agreement with the sub-contractor whereby the sub-contractor indemnifies the Contractor against loss or damage resulting from any breach of contract by the sub-contractor which in turn causes the Contractor to be in breach of his contract with the Employer. This is the basis of all the standard forms of sub-contract used in the construction industry.

The FCEC Form of Sub-Contract

The Federation of Civil Engineering Contractors has produced the Form of Sub-Contract, known as the FCEC blue form, for use with ICE 5 which, with very little modification, is suitable for use with the minor works form. The current version is the September 1984 revision.

The form is approved of by the Committee of the Associations of Specialist Engineering Contractors and the Federation of Specialist and Sub-contractors, and therefore is normally acceptable to representatives of both parties to the sub-contract. Copies of the form are available from the FCEC, and the ICE.

The few modifications that are necessary to enable the blue form to be used with the minor works form are given below. These modifications fall into two categories:

(i) those caused by differences in the definitions and drafting between the minor works form and ICE 5, and
(ii) the different methods of resolving disputes.

These modifications have been issued by the FCEC except for clauses marked with an asterisk which are are thought advisable by the authors.

Amendments to FCEC Form of Sub-Contract

For use with the minor works contract

Clause 1. (1) Add new sub-clauses:

(f) 'Constructional Plant' means all appliances or things required by the Sub-Contractor in the fulfilment of his obligations under the Sub-Contract.

(g) 'Temporary Works' means all temporary works required to be performed by the Sub-Contractor in the fulfilment of his obligations under the Sub-Contract.

(h) 'Section' means a Part of the Main Works as detailed in Item 7 of the Appendix to the Conditions of Contract of the Main Contract.

(i) 'Maintain' means the correction of defects as required by Clause 7 of the Main Contract.

Clause 2. (3) Line 2, after 'Sub-Contract Works' add a full stop, and delete the remainder of the clause.

Clause 3. (1) Line 6, delete 'Form of Tender' and replace with 'Conditions of Contract and Contract Schedule'.
Line 8, add 'for Minor Works' after the word 'Contract'.

Clause 4. (1) Delete whole sub-clause, and substitute 'Not used'.

Clause 5. (3) First line, delete 'Engineer's Representative' and substitute 'Resident Engineer'.

Clause 7. (1) Line 2, delete 'Engineer's Representative' and substitute 'Resident Engineer'.

Clause 13. (1) Line 6, delete 'and imperfection'.
Line 8, delete 'and imperfection'.

Clause 13. (2) Line 3, delete 'and imperfections'.
Line 6, delete 'and imperfection'.
Line 10, delete 'or imperfection'.

Clause 15. (6) Lines 4, 5 and 6, delete in toto and add 'Certifies the date on which the Contractor completed his obligations under the Main Contract'.

Clause 18. Delete whole clause and substitute

(1) If any dispute or difference of any kind whatsoever shall arise between the Contractor and the Sub-Contractor in connection with or arising out of the Sub-Contract, or the carrying out of the Sub-Contract Works (excluding a dispute concerning VAT but including a dispute as to any act or omission of the Engineer) whether arising during the progress of the Works or after their completion, it shall be settled in accordance with the following provisions:

(2) For the purposes of sub-clauses 18(3) to 18(5) inclusive and 18(8), a dispute is deemed to arise when one party serves on the other a notice in writing (herein called the Notice of Dispute) stating the nature of the dispute. Provided that no Notice of Dispute may be served unless the party wishing to do so has first taken any step or

invoked any procedure available elsewhere in the Sub-Contract in connection with the subject matter of such dispute, and the other party or the Engineer as the case may be has

(a) taken such step as may be required, or
(b) been allowed a reasonable time to take any such action.

(3) In relation to any dispute notified under sub-clause 18(2) and in respect of which no Notice to Refer under sub-clause 18(5) has been served, either party may within 28 days of the service of the Notice of Dispute give notice in writing under the Institution of Civil Engineers' Conciliation Procedure 1988 or any amendment or modification thereof being in force at the date of such notice and the dispute shall thereafter be referred and considered in accordance with the said Procedure.

(4) Where a dispute has been referred to a conciliator under the provisions of sub-clause 18(3) either party may within 28 days of the receipt of the conciliator's recommendation refer the dispute to arbitration if a person can be agreed upon by the parties by serving on the other party a written Notice to Refer. Where a written Notice to Refer is not served within the said period of 28 days, the recommendation of the conciliator shall be deemed to have been accepted in settlement of the dispute.

(5) Where a dispute has not been referred to a conciliator under the provisions of sub-clause 18(3) then either party may within 28 days of service of the Notice of Dispute under sub-clause 18(2) refer the dispute to the arbitration of a person to be agreed upon by the parties by serving on the other party a written Notice to Refer. Where a Notice to Refer is not served within the said period of 28 days the Notice of Dispute shall be deemed to have been withdrawn.

(6) If the parties fail to appoint an arbitrator within 28 days of either party serving on the other party a written Notice to Concur in the appointment of an arbitrator, the dispute shall be referred to a person to be appointed on the application of either party by the President (or if he is unable to act by any Vice President) for the time being of the Institution of Civil Engineers.

(7) Any such reference to arbitration shall be conducted in accordance with the Institution of Civil Engineers' Arbitration Procedure (1983) or any amendment or modification thereof being in force at the time of the appointment of the arbitrator and unless agreed in writing, shall follow the rules for the Short Procedure in Part F thereof. [* Such arbitrator shall have full power to open

up, review and revise any decision, instruction, direction, certificate or valuation of the Engineer or the Contractor.]

(8) If a matter which touches on or affects the Sub-Contract works is referred to a conciliator under the Main Contract the Sub-Contractor shall co-operate fully with the Contractor in the furtherance of the conciliation, and shall provided such information and attend such meetings as the Contractor may require in connection therewith. The Contractor shall inform the Sub-Contractor of any such conciliation, and the Sub-Contractor shall have the right to participate fully in it. The Sub-Contractor shall be bound by the recommendation of the conciliator to the extent that the Contractor is so bound.

* (9) If a dispute shall arise between the Contractor and the Sub-Contractor which arises out of the application, interpretation, or operation of the Main Contract, then the Sub-Contractor shall have the right to require the Contractor to refer the matter to a conciliator to be appointed under the Main Contract, and the Contractor will do everything necessary to obtain the appointment of a conciliator, and promote the cause of the Sub-Contractor. Provided always that conciliation is available under the terms of Clause 11.3 of the Main Contract.

Clause 19 Delete whole clause and substitute:

(1) The Sub-Contractor is deemed not to have allowed in the Price for any tax payable by him as a taxable person to the Commissioners of Customs and Excise.

(2) Tax invoices shall not be submitted to the Contractor by the Sub-Contractor.

(3) Where the Sub-Contractor is registered as a taxable person under the Value Added Tax Act 1983 as amended from time to time there shall be added to the amount of every payment made by the Contractor to the Sub-Contractor a separately identified amount equal to the amount of tax properly payable by the Sub-Contractor in respect of the taxable supply to which the payment made under Clause 15(3) relates.

(4) Where the Contractor operates the authenticated receipt system and upon receipt of the Contractor's Certificate of Payment and the payment detailed thereon the Sub-Contractor shall within 7 days return to the Contractor an authenticated receipt as required by the relevant Value Added Tax Regulations detailing the nett payment and the Value Added Tax on the said payment.

(5) Where the Contractor operates self-billing procedures no authenticated receipt will be required.

(6) Where the Contractor makes any provision pursuant to Clause 4 (Contractor's Facilities) and that provision is a taxable supply to the Sub-Contractor by the Contractor,

> the Sub-Contractor shall pay tax to the Contractor in accordance with the requirements of the Contractor at the rate properly payable by the Contractor to the Commissioners.

Notes on Clause 18

The wording of Clause 18 leaves the Contractor exposed if the sub-contract arbitrator should find that an Engineer's decision has been wrong and either no arbitration procedure has been started under the Main Contract or, if it has, the Main Contract arbitrator finds differently.

Since there is no power of joinder under the minor works form, the only way to avoid such exposure would seem to be to add a further sub-clause stating

*Clause 18. (10) If in the opinion of the Contractor the dispute concerns a matter in dispute or likely to be in dispute under the Main Contract then the dispute will not be referred to arbitration until such time as the matter is settled under the Main Contract. Provided that if the Contractor has not started arbitration proceedings under the Main Contract within a period of 60 days following a Notice to Refer from the Sub-Contractor the Sub-Contractor shall then be at liberty to re-issue the Notice to Concur and the provisions of this clause shall not then apply.

Completion of FCEC Form of Sub-Contract

Whilst it is possible to incorporate the terms of the FCEC blue form into the sub-contract by reference, it is recommended that the form should be completed in full for every sub-contract. When completed, the blue form will define the full extent of the sub-contract, and extreme care should be exercised in its completion. A few extra minutes spent at this stage may save considerable time, effort and cost at a later stage, should errors in it become apparent.

The following notes are intended to explain the intention of the various schedules, and to give guidance on their completion.

The parties (see Appendix E3)

On page 1, the date of the sub-contract and the names of the parties must be included. The date to be filled in should be the date upon which the last party to sign the Agreement signs it. This will normally be the Contractor. This date is relatively unimportant as the terms of the Agreement, when signed, will apply retrospectively to all works carried out in anticipation of it. (See legal comments at the end of this chapter).

The names of the parties must be the *correct* names of the parties. This may seem obvious but it is surprising how often errors occur in the names. This is particularly important for the Contractor, since it is essential that the name must be exactly the same as in the Main Contract. Generally, of course, this presents no problem. However, for groups of companies, a mistake can be made either in naming the parent company when it should be a subsidiary, or alternatively in naming the wrong subsidiary.

The address can be the local or head office of the company, or its registered address. Since the registered addresses will often be that of the company's solicitors or accountants, it is generally better to use the address of the office actually dealing with the Contract.

If the registered office address is given, the word 'of' should be deleted, and if another address is given, the words 'whose registered office is at' should be deleted.

Simple contract or contract under seal

The blue form is drafted on the assumption that it is be executed under seal, that is that it is a deed. Provision is made for the parties either to apply their common seals (usually an embossed impression of the company's name in a circle) or for them to sign the Contract alongside a small red stick-on seal.

This manner of execution will still suffice for corporations or companies (whether of the 'plc' or plain 'limited' varieties). However, for individuals the law has recently been changed so as to allow them to execute a deed without using a seal. All that is now necessary is for such parties to sign the Contract and for the words used in the 'formula of execution' to state clearly that it is being signed as a deed.

It is important for the Contractor that when the Main Contract has been executed as a deed, any sub-contract under it is also executed as a deed. Alternatively, the Contract can be executed as a simple contract,[1] in which case the normal means of execution, for corporations, companies and individuals alike, will be by signature alone.

Accordingly, on page 9 the passage beginning '*IN WITNESS whereof . . .*' should be replaced by

> *IN WITNESS whereof the parties hereto have caused this Agreement to be executed the day and year first above written.*

whether the Agreement is to be a simple contract or a deed. If it is to be a simple contract page 16 should be deleted and page 9 should continue thus

SIGNED by the above-named Contractor in the presence of:—

. . . .

. . . .

SIGNED by the above-named Sub-Contractor in the presence of:—

. . . .

. . . .

1. The essential difference between the two types of contract is the effect of the Limitations Act 1980, which limits the period during which an action can be initiated, either through the courts or in arbitration, after a breach has occurred. This period is six years for a simple contract, and twelve years for a deed.

Application

If, however, the Agreement is to be a deed, page 16 should be retained and completed. For corporations the first pair of alternatives on that page can be used as they stand, but for others the second pair of alternatives should be used with the addition of the words 'AS A DEED' after 'DELIVERED' in each case.

First schedule (see Appendix E3)

Contract details

The first schedule gives details of the Main Contract to which this is a sub-contract. The sub-contractor is deemed to have knowledge of the details of the main contract under Clause 3(1) of the sub-contract, so it is important that documents forming it are listed. Unfortunately, the printed form and the notes for guidance printed in the blue form at page 17 are incompatible, and the format in the notes for guidance should be followed in preference to the titles printed in the schedule.

Brief description of main Works

This defines the scope of the Works, and therefore the power of the Engineer to order variations. The easiest way to complete this part of the schedule is to copy the brief description of the Works, if there is one, from the Main Contract documents. If not, then one will have to be written. The description should describe the type and extent of the work to be performed.

Specified date

The specified date is the date when the Contractor will be submitting his monthly statement to the Employer in accordance with Clause 60(1) of ICE 5 and 6. Either specific dates or a statement such as 'first Monday in each month' is required.

Minimum amount of interim certificates

This figure is derived from Clause 60(2) of ICE 5 and 6, and the figure is found in the Appendix to the Form of Tender. It is not applicable after the issue of the completion certificate for the whole of the Works under the Main Contract.

The Contractor is entitled to withhold payment from the sub-contractor under Clause 15 (3)(b)(ii) if the Contractor's monthly statement does not come up to this amount.

Second schedule (see Appendix E3)

Further documents forming part of the sub-contract

This is probably the most important schedule, since it defines the scope of the sub-contract works, by listing all the documents that are to form the sub-contract.

The sub-contractor is deemed to have knowledge of the details of the Main Contract, but this is only for the purposes of evaluating damages in the event of a breach of sub-contract by the sub-contractor.

The requirements of the Main Contract are only included in the sub-contract to the extent that they are included in the documents forming the sub-contract. The second schedule performs the same function for the sub-contract as the contract schedule performs for the minor works contract itself. (see Chapter 3).

The documents to be listed may include

(i) sub-contractor's tender,
(ii) the Contractor's acceptance,
(iii) the Specification (or applicable extracts from the Main Contract specification),
(iv) the Drawings (numbers),
(v) conditions of the Main Contract that are to be specifically applicable to the sub-contract,
(vi) special conditions of the sub-contract $(1-x)$,
(vii) the Contractor's safety policy, and
(viii) correspondence.

These are dealt with below in turn.

Sub-contractor's tender or quotation

The sub-contractor's tender will often contain not only the rates and prices that the sub-contractor requires for the work but also special conditions that he attaches to his quotation. It may be, however, that certain of these conditions are unacceptable to the Contractor, particularly any 'standard terms of trading' which will generally be found in small print on the back of the quotation.

Standard conditions are automatically omitted by Clause 1(1)(b) unless they are separately specified in this schedule. Nevertheless, it is worth re-stating in the schedule as follows: 'The Sub-Contractor's quotation dated ..... but omitting the standard terms of trading printed on the reverse side.' It is often advisable if only the rates quoted by the sub-contractor in his tender are to be incorporated in the sub-contract to confirm that fact clearly in the schedule as a separate item as follows: 'The rates contained in the Sub-Contractor's quotation dated .....'.

Where there are considerable omissions from or changes to the sub-contractor's quotation negotiated prior to agreement, it is clearer and more convenient to extract these rates and prices and use them to form a new Schedule of Rates (or Bill of Quantities), and to list any conditions that are accepted in the special conditions. There is then no need to refer to the sub-contractor's quotation (or tender) at all, and the possibility of confusion is thereby avoided.

It should be remembered that the person negotiating the sub-contract is in the best position to be able to clarify and state exactly what has been agreed. It should not be left to the contract staff to try and decide from a whole heap of conflicting documents what the Agreement actually was.

Contractor's acceptance

If this letter is unconditional, it will bring the sub-contract into existence, and there is actually then no need for a formal sub-contract agreement.

Application

Generally, however, the letter will contain conditions additional to those contained in the sub-contractor's tender, and will therefore be a counter-offer, and not an acceptance. This counter-offer must then be accepted by the sub-contractor before a binding sub-contract is formed.

The tender and acceptance should only be included when subsequent modifications are not numerous. The alternative of extracting the information into a Bill of Quantities or Schedule of Rates, together with a list of special conditions, is normally to be preferred.

Specification

If only part of a standard specification is applicable, then the relevant clauses should be detailed. Sub-contractors often do not have large staff to read mountains of documents, and efficiency will be promoted if the amount of reading they have to contend with is kept to a minimum.

Drawings

The Main Contract drawings which are relevant to the sub-contract works, and upon which the sub-contractor's rates have been based, should be listed by number.

Conditions of the Main Contract

Any conditions of the Main Contract that are specifically applicable to the sub-contract works should be listed. Main Contract conditions are not automatically included in the sub-contract.

The list should include those clauses that deal with such matters as

(i) the provision of an agent,
(ii) providing watching and lighting,
(iii) provision of notices,
(iv) arrangements for termination,
(v) vesting of ownership in the Employer,
(vi) requirements for passes, access to site or photographs,
(vii) testing,
(viii) special hours of working, and
(ix) any special conditions in the Main Contract that are of particular application to the sub-contract.

Special conditions

Special conditions will cover any matters that have been specifically agreed between the Contractor and the sub-contractor, which are not contained in the Main Contract, such as

(i) normal site working hours,
(ii) safety provisions,
(iii) holiday arrangements,
(iv) arrangements for notices, additional information requirements and inspections,
(v) provisions for termination or omitting work from the sub-contract works if the sub-contractor is in default,

(vi) site facilities such as fences, lighting, watching, telephone, fax, temporary works, etc. to be provided by the sub-contractor (some of these may be for communal use, and this should be noted),

(vii) any arrangement with regard to material delivery and checking, and

(viii) resources to be provided by the sub-contractor.

It is convenient before the pre-contract meeting to list as special conditions all points that have been agreed in correspondence, and then to add to that list any points that may be agreed at the meeting. It is then a simple task to complete the sub-contract documentation.

Contractor's safety policy

It is a requirement of Section 2 of the Health and Safety at Work, etc. Act 1974, that all companies must prepare a written safety policy statement, and that this is communicated to their employees. This statement should be prepared by the Contractor and brought to the attention of all sub-contractors.

Correspondence

Letters passing between the parties which clarify the requirements of the sub-contract or detail points agreed subsequent to the sub-contractor's tender may be included specifically in the list of documents. However, care should be taken when doing so. Letters often contradict each other, raise points which are never answered, or modify previous agreements. If the correspondence becomes lengthy, it is often difficult without copious reading to determine afterwards what exactly was agreed.

The general rule should be that not more than four letters (and preferably only two), should be bound into a contract. If the correspondence exceeds this number, then the points agreed should be listed as special conditions.

The notes for guidance in the blue form on page 17 should also be read. Note particularly the importance of the second schedule when the work is priced on a lump sum basis.

'Sub-Contract Works'

This is a description of the type of work to be performed, and defines the limits of the work that the Contractor may order by variation under Clause 8. All the operations to be undertaken by the sub-contractor should be listed.

'Fluctuation Provisions (if any) affecting payments under this Agreement'

This item is to cover the cost of price increases experienced during the Contract. If it is left blank, then there are no fluctuation provisions, and the sub-contractor's rates and prices are not subject to adjustment as a result of increases in costs of supplying plant, materials or fuel.

This is generally the case under the minor works contract. (See Note 4 in the Notes for guidance). It is better, however, to write 'Fixed for the duration of the Contract', so there can be no doubt.

Nevertheless, if a fluctuation provision is to be incorporated, then it is suggested that this be limited to specific items, and a list of those items incorporated into the sub-contract documents. The Contractor should send out with the sub-contract enquiry documents a list of the items for which adjustment is to be made. The sub-contractor would then write against each item the cost for it upon which his offer is based. These costs should be supported by quotations, or other documentary evidence.

Third schedule (see Appendix E3)

Price

The price refers to a lump sum contract, where there is neither a Bill of Quantities nor a Schedule of Rates. In such a contract the sub-contractor will be expected to take-off the quantities himself and offer a lump sum for the work. Any discount for cash or main Contractor's discount should also be stated after the price. If such discount is described elsewhere in the documents (i.e. in the special conditions) a reference should be made to that clause so that it is clear that the discount can only be taken once.

If there are Bills of Quantities or a Schedule of Rates, no price should be entered and the words 'the price' should be deleted.

Retention

1. 'Percentage of Retention — Works'

This may be either a simple percentage, or a two stage percentage. (e.g. 10% on the first £20 000 and 5% thereafter). The blue form's notes for guidance recommend that a simple percentage should be used.

2. 'Percentage of Retention — Materials on Site'

This figure applies to amounts claimed in monthly statements for materials brought onto the Site for incorporation into the Works. Although the minor works form permits the inclusion of temporary works materials in the main Contractor's monthly statements the blue form does not provide the sub-contractor with a similar right. A simple percentage is appropriate.

3. 'Limit of Retention'

If there is to be a limit on the amount of the retention then an amount in pounds for the maximum sum to be retained is the simplest way of defining the limit. The use of a percentage of the price (which is one suggestion in the blue form's notes for guidance) only leaves the calculation for others later, and will not change the figure. When there is a Bill of Quantities or a Schedule of Rates a stated percentage is meaningless unless it is also stated to which figure (tender total) it is to be applied.

'Period for Completion'

The blue form's notes for guidance suggest that this should be a simple period of weeks or months. If there are separate sections for handover within differing periods, these should be stated here. All periods will be from the date of the start of the sub-contract works.

The required notice to be given by the Contractor to the sub-contractor to start the sub-contract work is stated in Clause 6(1) and is 10 days. The period for the sub-contract works starts either when the sub-contractor starts work on the Site or 10 days after the Contractor's written instruction is received by the sub-contractor, whichever is the earlier.

Any special programme conditions should also be stated here, such as:

(i) any additional notice required over the 10 days stated in Clause 6(1),
(ii) the number of visits provided for in the price, and the cost of any additional visits, and
(iii) any restraints imposed by the programme, such as 'Section 2 may not be commenced until 14 days after the completion of Section 1'.

Fourth schedule

Part I. Clause 4(2): Common facilities (see Appendix E3)

The fourth schedule details the facilities which will be provided by the Contractor to the sub-contractor, but which are not for his exclusive use.

The first section (A) is for construction plant. Craneage, compressors and pumps are obvious examples of plant that may be provided economically by the Contractor. The terms and conditions upon which each item of plant is to be provided are most important, as these will usually cover matters other than the price to be paid, such as

(i) availability (time, period, or convenience),
(ii) notice requirements prior to use,
(iii) any need for planning or other permission,
(iv) provision by a statutory undertaker,
(v) possibility of excess demand, or demand exceeding capacity,
(vi) use to be at the risk of the sub-contractor, who must indemnify the Contractor against all loss or damage caused by the sub-contractor, and
(viii) any restrictions on use.

If payment is required, the amount as well as the method for recording entitlement should be stated (e.g. times to be agreed, and recorded in writing at the completion of each use). The blue form's notes for guidance on page 18 should be read concerning payment terms.

Such facilities as water, electricity, car parking, canteen, and changing rooms which may be shared by both the Contractor's own employees and those of the sub-contractor, should be detailed in Section B.

Any restrictions on use, or conditions applicable to the use of these facilities (as given above under 'Further documents forming part of the sub-contract') should be stated.

Part II. Clause 4(2): Exclusive facilities (see Appendix E3)

As the title says, this part details any facilities which may be made available for the exclusive use of the sub-contractor.

All the comments on Part I apply to this part. However, it may be useful to include items of plant and equipment which the Contractor is likely to have on the Site and which may be of use to the sub-contractor, so that a rate is decided and agreed in advance of use. This will prevent unnecessary argument on Site. If this is done, it must be made clear that the provision of such plant is subject to availability and at the sole discretion of the Contractor.

Fifth schedule (see Appendix E3)

Part I. Sub-contractor's insurances

The notes for guidance on page 18 of the blue form should be read 'Part 1 details the insurance policies that the Contractor requires the Sub-Contractor to provide.'

The Contractor must ensure that all the insurances required by the sub-contract are covered either by himself or by his sub-contractor. It must be remembered that an insurance policy indemnifies the policy-holder against a liability, and sub-contractors are only covered by the Contractor's policy if they are specifically made a party to it (that is if they are made 'joint assured').

If the Contractor's policy does not extend to sub-contractors, and an event occurs for which the sub-contractor is responsible, the Contractor may be unable to claim for any damage he suffers under his policy of insurance. If the sub-contractor is not insured for it, and the damage is more than the sub-contractor can pay, the Contractor may be left having to bear the loss himself.

Main Contractors must ensure that they are covered in respect of their sub-contractors' actions for all matters for which they are liable under the Main Contract. Thus, if the Contractor is responsible for insuring the Works, either he or the sub-contractor must take out insurance policies to cover the sub-contractor's liability should the Works or the Contractor's plant be damaged.

It is, however, not an all-or-nothing-at-all situation. The Contractor may take out full cover on behalf of himself and his sub-contractors, but need not pass on the total benefit to the sub-contractor. For instance, the Contractor will insure the construction plant and personal equipment and effects of his own employees while still requiring the sub-contractor to cover these items on behalf of the sub-contractor's employees. This is because the cost of processing such claims can be expensive, and they are matters over which he has no control. To require the sub-contractor to cover these items also provides an incentive to the sub-contractor's management to safeguard his own plant and ensure that his employees likewise protect their own personal belongings.

The effect may be that such matters are doubly insured, but compensation can only be claimed once, since the sub-contractor has no right to claim against the Contractor under the sub-contract.

The matters for the Contractor to consider when negotiating the sub-contract are:

(i) loss or damage to the Works,

(ii) third party liability,
(iii) constructional plant and equipment,
(iv) employer's liability insurance (required by law),
(v) professional indemnity (if the sub-contractor has a design responsibility),
(vi) personal property of the sub-contractor, and
(vii) personal effects of the sub-contractor's employees.

If the sub-contractor is to insure the works, such policy should be in the joint names of the Contractor, the Employer and the sub-contractor.

Part II. Contractor's policy of insurance

This part details the insurance policies taken out by the Contractor, and of which the sub-contractor is to have the benefit under Clause 14(2). It is important that the Contractor should state clearly the total sum of the insurance policy, and what matters are covered in it, such as

(i) the Contract Works, both temporary and permanent,
(ii) materials and equipment brought to the Site for the purposes of the Contract,
(iii) professional fees,
(iv) transport to and from the Site,
(v) removal of arisings,
(vi) plant and equipment (owned or hired),
(vii) temporary buildings and their contents,
(viii) personal effects, and
(ix) transport of money.

Any conditions applicable to the policy, or excesses provided for, should also be specified.

Where there are excesses, a statement concerning who is to meet these excesses in the event of a claim should be either written into Part II, or included in the special conditions.

Normally, the sub-contractor will be expected to bear the cost of such excesses in respect of claims that the sub-contractor makes. If these excesses are considered too large, the sub-contractor can always take out a separate policy to cover them in whole or in part.

Chapter 18

THE UNFORESEEN

Introduction

In all civil engineering contracts the Contractor undertakes to carry out the Works specified in the Contract within the time stipulated in the Contract, and the Employer undertakes to pay the Contractor for those Works in the manner set out in the Contract. If everything turns out to be as both the Employer and the Contractor expected when they entered into the Contract all should be well. However, it is inherent in works of civil engineering construction that the actual situations encountered during their construction may be very different from those envisaged at the time of tender.

The legal relationship between the parties to a contract are defined by the Contract itself, as are their respective rights and obligations under it. It follows, in principle, that only those matters which have been foreseen by the parties at the time the Contract was entered into will be covered by the provisions of the Contract. Thus, again in principle, unforeseen situations will fall outside the scope of the Contract. This is not to say that the Contractor will be relieved from his obligations to complete the Works, but that any necessary work undertaken by him and not envisaged at the time of the formation of the Contract will not entitle him to payment at the rates in the Contract but to a reasonable and fair payment from the Employer (that is, the Contractor will be entitled to payment on a *quantum meruit* basis).

Unforeseen situations may also prevent the Contractor from completing the Works within the time required by the Contract and, to that extent, the Contractor will be in breach of contract although he is not himself at fault. The Contractor's failure to complete as required will, again in principle, entitle the Employer to terminate the Contract. However, such an action is unlikely to assist the Employer in getting the project finished. If, on the other hand, the Employer condones the Contractor's breach by allowing him to continue, the time provisions in the Contract will be nullified and the Contractor will only be bound thereafter to complete the Works within such time as is 'reasonable' in all the circumstances. In practice, that means 'when he gets round to it' provided, of course, that such completion is not visibly protracted. Moreover, the Employer will not only have lost any definite date for completion, but may also be deemed to have waived any right to liquidated damages for delay.

Due to the high risk of the unforeseen occurring on a civil engineering contract, this possibility of having both the completion date and the ultimate cost of the Works rendered uncertain is clearly unsatisfactory to both the Contractor and the Employer. This uncertainty can to some extent be reduced by including in the Contract provisions and procedures for dealing with unforeseen events as they arise.

Thus, although the events themselves may be unforeseen and outside the expectations of the parties when they entered into the Contract, the means of dealing with them will have been foreseen, and, to that extent, the resolution of such problems (as distinct from the problems themselves) will have been brought within the scope of the Contract.

Unforeseen events

For contracts under the minor works form unexpected situations may arise from actions of the Engineer, actions of the Contractor or events outside the control of either party.

Actions of the Engineer

The Engineer is given very wide powers, under the minor works form, to give instructions and directions to the Contractor. Such instructions may arise either because the Contractor has not complied with his obligations (in which case the Contractor is in default) or from a desire by the Engineer to change what the Contractor is required to do. The former will not entitle the Contractor to any additional time or money whereas the latter may do so, depending upon its effects on the operations of the Contractor.

When faced with a demand for extra payment by the Contractor, the question for the Engineer is whether or not the instruction resulted from some default on the part of the Contractor. For instance, there are detailing errors on most contracts, but the Contractor cannot be expected to include for them in his price because they involve no default on his part. Thus errors in reinforcement, inability to pour concrete because of the closeness of the bars, laps in reinforcement which are impossible to make because of the density of the steel, and inserts which are impossible to place because they clash with the reinforcement all occur regularly. The type of error is clearly foreseeable, but the actual error is not. They are therefore unforeseen when they occur. None of them is a matter for which the Contractor can be expected to allow.

Although not strictly an action of the Engineer, because they occur before the Contract is formed, errors that are found in the Contract Documents will normally be the responsibility of the Employer.

Actions of the Contractor

The Contractor may decide to change his methods of construction, the order in which he intends to carry out the Works, the sources of the materials which he intends to use and the staff he deploys upon the Site. These changes may result from the fact that his original proposals were impossible or uneconomical to carry out.

All such actions are within the control of the Contractor and do not give rise to any claim or release the Contractor from his obligations under the Contract. But nor do they normally give rise to claims by the Employer for any additional costs that may be caused as a result of abortive work undertaken by the Engineer. The minor works contract does not require the Engineer to agree or check the Contractor's proposals so no such expenditure is involved unless the Specification stipulates that the Contractor must obtain the Engineer's approval to his methods of construction and/or the design of his Temporary Works.

Only if a method or sequence of working is specified in the Contract and then proves impossible does it give rise to any consideration under the Contract, and then the Engineer must order a variation to overcome the impossibility.

Events outside the control of either party

All hazards that increase the costs of a project are the responsibility of one or both of the parties, and the Contract determines which one. Liability is not affected if one or other of the parties is insured because insurance is only payable by the insurance company to the party who has incurred a liability under the Contract. If there is no liability then no insurance money will be forthcoming.

Unforeseen events outside the control of either party can be divided into those occurring outside the Site and those occurring within the confines of the Site.

1. Outside the Site

Events affecting works outside the confines of the Site will include

Excepted Risks:

(i) war,
(ii) civil war,
(iii) rebellion or revolution,
(iv) nuclear explosion, or radiation, and
(v) things falling from aircraft.

The Excepted Risks are often uninsurable. Contractually they are taken by the Employer.

Other risks:

(vi) strikes or lockouts (other than those affecting the Contractor's employees,
(vii) damage or explosion at works from which materials may come,
(viii) actions of overseas governments (such as increasing taxes or imposing export bans,
(ix) new legislation by the UK government and tax changes,
(x) changes in planning laws,
(xi) restrictions on access to the Site,
(xii) weather, floods, excessive heat or snow conditions,
(xiii) shortages of labour, materials or plant, and
(xiv) increases in bank rate, VAT or company taxes.

In general, the costs incurred as a result of all these 'other risks' are the responsibility of the Contractor unless they force a change in the design, or the Contract provides otherwise, when they would become the responsibility of the Employer. However, if they result in a delay to the completion of the

Works then the Contractor will probably be entitled to an extension of time.

Under the minor works form the Contractor may be entitled to extensions of time for completion under Clause 4.4. for most of these risks and possibly to additional payment under Clause 6.1 of an amount that the Engineer considers fair and reasonable.

2. *Inside the confines of the Site*

The most common unforeseen situation that occurs on the Site is discovering, as the excavation proceeds, that the actual ground conditions are significantly different from what was expected. There are, however, others such as the non-availability of parts of the Site, the condition of foundations to adjacent properties, ingress of deleterious matter from outside, flooding and trespassers, all of which can affect the progress of the Works.

Providing for the unforeseen

Almost all of these unforeseen situations are covered to some extent by the provisions of the minor works form, and can be conveniently considered under the following headings.

1. *Changes to the Works*

Changes to the Works to be undertaken by the Contractor can arise in two ways

(i) the Engineer may order variations, or give other instructions under Clause 2.3, or
(ii) the Works may be altered as a result of encountering adverse physical conditions or artificial obstructions pursuant to Clause 3.8.

Engineer's variations may be prompted by the need to improve the functioning of the Works or they may be necessary to make the Works capable of completion. Such variations are 'unforeseen' at least in the sense that if the need for the variation had been realized at tender stage then, no doubt, the tender documents would have included it.

Other matters, such as the ordering of tests, changing the sequence of works, and ordering the Contractor to provide services to other contractors, will fall into the same category. Nevertheless, most of such changes to the work undertaken by the Contractor will be within the general scope of the parties' intentions.

Such changes may, in appropriate circumstances, entitle the Contractor to additional payments (or the Engineer to make deductions) under Clauses 6.1 and 6.2. and are discussed fully in Chapter 9. They may also give rise, where appropriate, to an entitlement to an extension of time under Clause 4.4. Again this is discussed fully in Chapter 7.

2. *Errors in the contract documents*

Any error that is found in the contract documents will be 'unforeseen' in the sense that had it been discovered prior to tender doubtless the documents would have been corrected. The Engineer is given power under Clause 2.3(g) to 'elucidate or explain any matter to enable the Contractor to meet his obligations under the Contract'. That will include errors in the contract documents.

At the time of tendering the Contractor must make various assumptions based upon the quantities and types of materials that will be required, and will base his rates upon those assumptions. Clearly, such assumptions will rarely turn out to be entirely accurate, but minor deviations will be of little consequence either because the amount to be paid to the Contractor under a 'measure and value basis' will be adjusted automatically, or because under a 'lump sum' contract these variations will be *de minimis* and may also be amenable to a certain amount of 'swings and roundabouts'.

However, if the quantities are found to differ substantially from those in the tender documents, or large items are found to have been omitted altogether, it can seriously affect the validity of the tender assumptions, and cause great difficulty either for the Contractor (because his overheads have been spread over a higher quantity than is subsequently measured) or for the Employer (because he will have lost any 'benefits of scale' which would have affected the rates and prices had the proper quantity been known). Accordingly, the minor works form permits the Engineer to adjust the rates under Clause 6.1 if the Contractor incurs any additional cost, or reduce them if any work is omitted. In each case the test to be applied is what is 'fair and reasonable' in the circumstances.

The equivalent clauses in both ICE 5 and 6 are to be found in Clause 52(2), and 56(2) in the case of changed quantities, Clause 55(2) for omitted items, and Clause 5 for errors in the documents on other matters.

3. Excepted Risks

The Contractor is relieved by Clause 3.3(1) of his liability for damage to the Works arising from any of the Excepted Risks (as defined in Clause 1.5). Since these risks are taken by the Employer, it follows that that the Contractor would be able to recover any additional cost to him resulting from any delay caused to the completion of the Works, and he would be entitled to an extension of time for that delay under Clause 4.4 if it is necessary to avoid the payment of liquidated damages for late completion. (The effects of delay are discussed below.) The Contract provides that the Engineer may order the Contractor to make good such damage at the expense of the Employer.

4. Ground conditions

Unlike Engineer's variations, the effects of adverse physical conditions and artificial obstructions will always be unforeseen in the true sense of the word, since, at least, the Contractor's intended sequence and/or methods of construction may be either disrupted or changed entirely in order to overcome the problem. In either case the Contractor will probably be delayed and put to additional cost.

The minor works form, like ICE 5, ICE 6 and GC/Works/1, has adopted the policy that unforeseen ground conditions or artificial obstructions should be the responsibility of the Employer. Clause 3.8 relieves the Contractor from the additional costs arising from any delay, disruption or changed methods experienced in overcoming the problem and the Engineer is given power under Clause 2.3(e) to give instructions for measures to deal with such conditions or obstructions.

There is a wide range of possible changes in the ground conditions that may cause difficulty to the Contractor when excavating or constructing the Works below ground. These may include the following:

(i) The excavated material may be either harder or softer than expected.
(ii) The ground water may either be at a higher or lower level than anticipated, or it may be different from the predicted quantities due to increased or decreased permeability.
(iii) Geological faults may be experienced which require special methods of treatment.
(iv) Subterranean caves or cavities may be found which require filling.
(v) Subsidence due to mining or for any other reason may require special ground treatment or even prompt a change in the design of the Works.
(vi) Exploitable minerals may be encountered, such as coal, ore, oil or gas, which delay the Works.
(vii) Dangerous gases, such as methane, black damp, or fire damp may be found in the excavations which require special precautions to be taken to avoid explosions, or other danger.
(viii) An earthquake is a physical condition. The effect of the earthquake could require extensive remedial work. Although these effects are normally covered by insurance the Contractor would be able to recover any additional costs under the Contract. It would be for the Employer to make the insurance claim.
(ix) The finding of fossils and antiquities is not covered by the minor works form, but such findings would qualify as artificial obstructions. Any consequential disruption or expense would be recoverable by the Contractor.
(x) Underground services, which are uncharted and undetected will be classed as unforeseen.
(xi) Tunnels and shafts from old mine workings or other underground workings. Very few of the early mine workings are charted with any accuracy.
(xii) Old roads, foundations, basements, and other man-made structures may be discovered.
(xiii) Burial grounds, historical sites, or plague pits may be found.
(xiv) Unexploded bombs may be found.
(xv) Dumped hazardous materials requiring special disposal methods may be found.

When considering claims for additional cost or extensions of time in respect of any of these items the Engineer must first decide whether or not what was found can be classified as being something which 'could not reasonably have been foreseen by an experienced contractor', and only if the answer is 'yes' does the Contractor's claim merit further consideration.

Of course, any of the above items could in the very broad sense be foreseen because they are all matters with which most people have either heard of or have actually seen at some time during their lives. However, the policy behind Clause 3.8 is that the Contractor should not allow in his tender price for matters that could not reasonably have been foreseen, and it must be a fact that at least nine contracts out of ten do not experience most of the matters covered above. It therefore follows that the Contractor will not be expected to cover for dealing with them in his tendered price unless there is some evidence that they are likely to be encountered. The Engineer, when rejecting a claim under Clause 3.8, should therefore always be prepared to say exactly why he thinks that the Contractor should have expected to find on the Site what he did in fact find. This will usually require some degree of physical evidence.

One special category of adverse physical condition is weather or conditions caused by weather. The policy is that the Contractor shall always be responsible for costs incurred by him as a result of weather conditions because he is the party best able to control those costs, but that he will be entitled to an extension of time in respect of exceptionally adverse weather if it is needed to avoid payment of liquidated damages for late completion.

5. Situations that cause delay

The Contractor is not only obliged to complete the Works within the time laid down in the Contract, but is also entitled to the use of all of that time if he so desires. It follows that any 'act of prevention' on the part of the Employer will entitle the Contractor to an extension of time equivalent to that lost through the Employer's act of prevention, whether he needs it or not.

On the other hand there is no presumption that the Employer (or his agent, the Engineer) will so arrange their affairs as to facilitate the Contractor's preferred methods or timing of construction [*Glenlion Construction Ltd* v *The Guinness Trust* (1987) 30 BLR 89]. Such acts of prevention apart, there will be many other potential situations which will cause the Contractor delay, but in general they will only entitle the Contractor to extensions of time if he needs them in order to avoid a breach of his obligation to complete on time.

Assessing the effects of delay and disruption

The effect of the unforeseen is generally to cause delay and disruption to the Contractor's operations. It is easiest to consider delay as a matter concerning the Contractor's overheads and disruption as affecting the operations in the field. A delay will occur when the date upon which the operation or the Contract could have been completed, but for the finding of the physical condition or artificial obstruction, is put back. Disruption occurs when the resources committed to the operation are adversely affected, either because the outputs which could be achieved but for the obstruction are reduced, or because more (or different) resources have to be deployed.

Delay is time-related, whereas disruption is production-related.

Delay

Delay may affect either the whole of the Contract or just a single operation. In both cases, the measure of the delay is the amount by which the date for completion is put back as a result of the condition or obstruction.

The measure of cost for a delay to the whole of the Works will be the total of the Contractor's overhead running at the time that the delay takes place. For a delay to a single operation which does not affect other operations, the delay costs will simply be the overhead costs specifically associated with that operation.

Contractor's overheads during the period of such delay may be regarded as 'thrown away', that is expended needlessly, and are therefore recoverable if the delay is from a cause for which the Employer is responsible (see Clause 6.1).

The Contractor's site overheads will comprise the following:

(i) Staff costs.
(ii) Staff on-costs, such as pensions, cars, holidays, bonus and National Insurance contributions.
(iii) Provision and running of offices.
(iv) Consumables, telephone, fax, printing.
(v) Equipment.
(vi) Insurances.
(vii) General site labour for non-measured items, such as unloading vehicles, cleaning the Site, putting up warning signs.
(viii) Plant whose time on the Site is not dependent upon the work it does, but upon the time for which it is needed (generally defined as overhead plant). Such items will be general purpose craneage, bar benders and croppers, general site transport, compressors, pumps and workshop equipment.
(ix) Costs of the Resident Engineer. If these items are included in the Contractor's rates, and are not recoverable on a measured basis, they will be considered an overhead cost.

Contractors usually keep accurate records of overhead costs, so the figures should not be difficult to obtain.

The term 'cost' includes overhead costs both on and off the Site under the minor works form (see Clause 1.3). The Contractor's costs off the Site are for his head office, and for any workshops or offices involved in the Contract.

It is very difficult to identify actual costs expended by the Contractor's head office organization upon one particular contract out of the many that are concurrently being undertaken. As a result a number of methods of assessing them have been devised, and some have had judicial recognition.

The total head office costs of a contractor may be expressed as a percentage of the annual turnover of that contractor. Most contractor's head office costs are in the 3%–7% range, depending upon how certain costs, such as the Contract Manager, are allocated between head office and site. Contractors will make an allowance in their tender costs for their particular percentage.

The easiest method of allowing for head office costs in assessing costs for delay is therefore to apply the percentage allowance that the Contractor used in his tender to the total of the other additional costs that were incurred. This figure can be obtained from the Contractor. The principle behind this method is that the additional costs are part of the Contractor's turnover and should therefore attract the same overhead expenditure.

Application

More complex methods have been evolved of which the Hudson formula, so named after the legal textbook in which it was first suggested,[1] is the most commonly used. This involves multiplying the percentage allowance made by the Contractor in his tender by the contract sum, and dividing the result by the Period for Completion (as defined in the Appendix Item 6) to produce a weekly allowance for head office. This figure is then applied to the period of the delay to give a recoverable cost.

There are a number of objections to the use of the Hudson formula. First, it is to be used only when the Contract period is exceeded, and secondly, by using the contract sum as the multipicand it uses a figure which already includes the head office allowance. An element of the cost is therefore included twice.

Refinements on the Hudson formula are the Ebden and Eichleay formulae, which, instead of relying on the Contractor's tender allowance, use instead the Contractor's actual head office cost, as certified by their auditor, and divide it by their actual turnover at the time of the delay. There may be an application for such refinements on major contracts, but for minor works the simple percentage addition to the other costs would seem most appropriate.

When the delay is to a single operation, and not to the whole of the Works, then only the overheads directly attributable to that operation are recoverable. These may include

(i) a section engineer, foremen and other staff engaged upon that operation, and any accommodation that may be required by them;
(ii) standing plant such as compressors, cranes, scaffolding, shutters; and
(iii) labour, and other resources that cannot be found alternative work.

It will be for the Contractor to identify which overhead costs are additionally incurred as a result of the delay. However, if the percentage addition method is used for evaluating the head office costs of any delay, it will be applicable delays to a single operation. The formula methods should not be used in this way.

Disruption

At the time of tender, the Contractor's estimator will build up his rates by calculating the cost of the gang necessary to undertake an operation and dividing that cost by an assumed rate of output to obtain a unit cost. For instance, an excavation gang of two men and a JCB may be expected to turn out 8 cubic metres of excavation per hour. If the cost of the JCB and the two men is £36.00 per hour, then the cost of excavation per cubic metre is £4.50. If, as a result of an adverse physical condition, the rate of excavation is reduced to 6 cubic metres per hour then the cost will go up to £6.00 per cubic metre and the measure of the disruption is £1.50 per cubic metre.

Under the minor works form the Contractor recovers the cost resulting from such delay and disruption as may be incurred as a consequence of Engineer's instructions, under Clauses 6.1 and 4.4.

---

1. *Hudson's building contracts*, 10th edition (1979), p. 599, London: Sweet & Maxwell.

If the Contractor has to do additional work or use additional plant or equipment (that is work, plant or equipment that would not have been used if the condition or obstruction had not been encountered) then the cost of so doing is recovered under Clause 3.8. By definition the gang and the equipment that was being used on the operation when the condition or obstruction was encountered cannot be additional, but any increased unit cost due to less productive working will be covered by a disruption claim.

Comparison with ICE 5 and 6

Although the provisions vary somewhat between the three forms the principles behind all three are identical. Many of the clause comparisons have been mentioned already in the text, but there is one detail difference that is worth noting.

Under both the minor works form and ICE 6, the Contractor is obliged to give the Engineer notice whenever he finds an unforeseen physical condition or artificial obstruction whether or not he intends to claim additional payment or use extension of time. This notice, which is not required under ICE 5, is to enable the Engineer to assess whether or not there will be any effect upon the design resulting from the changed condition. It will also enable the Engineer to initiate record-keeping if there is any likelihood of a claim being advanced by the Contractor at a later date.

Legal comment

Normally, only the parties to a contract can amend it. However the minor works form, in common with the other ICE forms and, indeed, most standard forms of contract in the construction industry, gives express power to a person who is not a party to the Contract — namely the Engineer — in effect to re-write the Contract as necessary in the face of unforeseen situations. This is a unique provision, and is matched by an express power in the arbitrator (should a dispute go that far) to exercise this amending power by 'opening up, reviewing and revising' any decision or other action of the Engineer. Such a wide-ranging power has always been outwith the prerogative of the courts and can only be exercised by the Engineer or arbitrator as the case may be. However, a court may now be expressly empowered, by reason of the Courts and Legal Services Act 1990, to exercise such jurisdiction, but only if all parties to the action expressly agree.

Taken as a whole, these provisions preserve the existence and integrity of the Contract while allowing it to be adapted to meet each unforeseen situation as it arises. While neither the situations nor the solution to the resulting problems could be predicted, the means whereby the time and payment provisions of the Contract are to be adjusted are expressly set out in the Contract, thereby preserving the balance of advantage between the parties and maintaining the desirable provisions (which would otherwise be lost if the Contract was a nullity) while sensibly regulating the adaptation or replacement of those now seen to be inappropriate.

In addition, the provisions on extensions of time prevent time from becoming 'at large' and preserve time as being 'of the essence', so that the Employer still knows when he will obtain possession of the completed Works (albeit later than originally envisaged) and the Contractor is still set a target date upon which to complete. The Employer's right to liquidated damages is also preserved.

The provisions of the Contract thus ameliorate to some extent the harsh doctrine of frustration, whereby the Contract may be considered at an end if it became impossible to perform. Whilst the effect of this principle is usually uncertain when applied to any particular situation, what is quite certain is that it would cost one or both of the parties a considerable sum of money in legal expenses to establish their respective rights precisely. It is clearly better that these risks be ascribed to one or other of the parties so that that party can take the particular risk into account before deciding to embark upon the Contract.

Whether or not a given situation could have been foreseen by an experienced contractor is a matter of fact, not law, and as a result of the *Northern Regional Health Authority* v *Derek Crouch* (1984) and the 1979 Arbitration Act, it is unlikely that there will be much further guidance on this subject. In any event, there have been remarkably few cases covering unforeseen circumstances. In *Appleby* v *Myers* (1867) it was decided that the Employer did not warrant the Site fit for the Works, which means that the Contractor is responsible for unforeseen conditions or obstructions unless the Contract provides otherwise. The minor works form in Clause 3.8 provides for the Employer to pay such additional costs.

The main cause of argument is usually about what a Contractor could reasonably have foreseen. In *C. J. Pearce and Co.* v *Hereford Corporation* (1967), a contract under ICE 5, the Contractor was engaged to construct a new sewer under an existing old one. The sewer collapsed when the ground was disturbed around it. It was held that the Contractor could have foreseen that an old sewer might collapse when a new one was driven under it, and the Clause 12 claim was therefore rejected.

Clause 12 is similar to Clause 3.8 of the minor works form and the judgment would seem to apply equally to both forms. However, it does appear that this judgment had more to do with the care to be exercised by the Contractor when actually executing the work that to foreseeability. It is, still, the only case with any relevance to foreseeability.

In interpreting what is covered by Clause 3.8, reference must be made to the whole Contract, and in particular the Specification, Bills of Quantities and the Contract Drawings. It is the intention of the Contract that is important, and it is clear from *Holland Dredging (UK)* v *The Dredging and Construction Company Limited and ICI* (1987) that there is no requirement for any supervening event, which means that Clause 12 and Clause 3.8 will be valid whether the physical condition or artificial obstruction complained of existed at the time of tender or arose at a later date.

Chapter 19

ICE CONCILIATION PROCEDURE

Introduction

The ICE Conciliation Procedure which accompanies the Conditions of Contract for Minor Works was the first standard conciliation procedure to be published for use in the construction industry in the UK.

Both ICE 5 and the minor works form provide for the resolution of disputes prior to arbitration. Experience over many years has shown that such a safety valve is valuable, firstly in that it provides a pause for second thoughts on the matter in dispute, and secondly in that the Engineer's formal decision (under ICE 5) or conciliator's recommendation (under the minor works form) can often provide a basis upon which negotiation can recommence and an agreed settlement be achieved. Lastly, even if no such settlement can be reached, the exercise itself will often clarify the real issues in the parties' minds and indicate the relative strengths and weaknesses of their several positions so that, where arbitration becomes inevitable, the conduct of the reference will be more efficient and thus both quicker and cheaper.

Reference back to the Engineer under ICE 5, Clause 66(1) is compulsory in that it is a condition precedent to arbitration. Conciliation under the minor works form, however, is voluntary, and it is open to either party to pre-empt the other by requiring the dispute to go direct to arbitration. On the other hand, a party wishing to resort to conciliation must first invoke any other procedure (other than arbitration) which may be available to them under the Contract, a proviso which does not apply under ICE 5. Those differences apart, however, the *results* obtained under ICE 5, Clause 66(1) and conciliation respectively are similar in that both the Engineer's Clause 66(1) decision and the conciliator's recommendation will become final and binding upon the parties unless challenged within the time limits set down in the Contract by service of a notice to refer the dispute to arbitration.

ICE Conciliation Procedure 1988

The ICE Conciliation Procedure has been agreed between the Institution of Civil Engineers, the Association of Consulting Engineers and the Federation of Civil Engineering Contractors for use with the ICE Conditions of Contract for Minor Works. It is, however, available for use under any other form of contract which provides for conciliation, and may also be used with advantage where there is no such contractual provision but the parties to a dispute wish to adopt some form of dispute settlement which is less formal (and thus, hopefully, both quicker and cheaper) than litigation or arbitration.

Application

Unless otherwise stated, references in this chapter are to the rules in the ICE Conciliation Procedure 1988.

Agreement to conciliation

> Rule 1. These rules are deemed to have been agreed between the parties to apply whenever they have entered into a contract which provides for conciliation for any dispute or difference which may arise between the parties in accordance with the Institution of Civil Engineers' Conciliation Procedure 1988 or where the parties agree the conciliation will apply in accordance with that procedure.

In the case of the minor works form, the necessary provision for conciliation is contained in Clause 11.3.

Where a contract contains no such provision, or where, in the absence of any contract, the parties agree to conciliation, it would be prudent to record the agreement in writing, even if only in an appropriate exchange of letters. Formal words are not necessary, but the meaning must be clear and unambiguous. One possible example might be: 'we the undersigned hereby agree to refer the following disputes to conciliation in accordance with the ICE Conciliation Procedure 1988' followed by a sufficient description of the disputes to be referred to identify the subject matter.

Interpretation

> Rule 2. These rules shall be interpreted and applied in the manner most conducive to the efficient conduct of the proceedings with the primary objective of obtaining the conciliator's recommendation as quickly as possible.

The aim of conciliation is to produce a quick result and to avoid the misuse of the procedure as a means of delaying a decision. Since a conciliator is free from most of the restraints imposed upon an arbitrator by law, he is able to proceed in any manner that he thinks fit.

Notice of conciliation

> Rule 3. Subject to the provisions of the Contract relating to conciliation until such time as the Contractor has completed all his obligations under the Contract either party to the Contract may by Notice in writing to the other party or parties request that any dispute, difference or other matter in connection with or arising out of the Contract or the carrying out of the Works shall be referred to a conciliator for his recommendation. Such Notice shall be accompanied by a brief statement of the matter or matters upon which it is desired to receive the conciliator's recommendation and the relief and remedy sought.

> Subject to the provisions of the Contract relating to conciliation until such time as the Contractor has completed all his obligations under the Contract . . .

There is no provision in the minor works form that conciliation must await the completion of the Works under the Contract. The only restraint to referring a dispute to conciliation is that all the steps that are available under the Contract must have been taken, or that the other party has had a reasonable amount of time to take such steps (Clause 11.2). For instance, if the Contractor wishes to claim an additional amount in relation to an instruction given by the Engineer, Clause 6.1 requires that there be consultation between the Contractor and the Engineer on the amount to be paid. It will be necessary for a party to have requested such consultation prior to being able to refer the evaluation to a conciliator.

> . . . either party to the Contract may by Notice in writing . . .

All that is necessary to start the conciliation process going is for one party to serve a notice upon the other requesting that the dispute or difference shall be referred to a conciliator. Such notice should usually have been preceded by a discussion as, since conciliation is a voluntary procedure, it is unlikely to be aided by a pre-emptory notice coming out of the blue. The notice will usually follow a meeting at which the Engineer and the Contractor have been unable to reach agreement and have agreed that the easiest way out of the dilemma is to ask a conciliator for a recommendation.

> Such Notice shall be accompanied by a brief statement of the matter or matters upon which it is desired to receive the conciliator's recommendation . . .

This rather formal wording should not be taken to mean that this brief statement must be a formal document. The intention is that it should set out the subjects which the conciliator will be asked to consider simply and concisely as an introduction to the dispute for the conciliator when he first takes up his appointment.

> . . . and the relief and remedy sought.

Again, there should be nothing in the reliefs or remedies that are unfamiliar to the recipient party, since such matters will already have been discussed. They do, however, set the boundaries of what is required, and avoid the recipient party agreeing to conciliation in circumstances where he would not be prepared to compromise on the matters in dispute.

Appointment of conciliator

> Rule 4. Save where a conciliator has already been appointed, the parties should agree a conciliator within 14 days of the Notice being given under Rule 3. In default of agreement, any party may request the President (or, if he is unable to act, any Vice President) for the time being of the Institution of Civil Engineers to appoint a conciliator within 14 days of receipt of the request by him which request shall be accompanied by a copy of the Notice.

The essential feature of conciliation is that it is a voluntary process to be undertaken by agreement. It is therefore preferable that the parties should themselves choose the conciliator. Clearly, if they can select somebody who is known to them both, and whose opinions and judgement they trust, then that is the ideal solution. But, if not, the parties can always refer to the lists maintained by the Institution of Civil Engineers of persons who are suitable and willing to act as conciliators. These lists are regional, so as to minimize any travelling expenses that may be involved for the conciliator.

If, having studied the appropriate lists, the parties are still unable to agree, application can then be made to the President of the Institution for him to appoint a conciliator. Such an application cannot be made until at least 14 days after the original notice of conciliation under Rule 3 has been given but, once an application has been made, the President (or a Vice President acting on his behalf) will normally make an appointment within a further 14 days. It is for this reason that Rule 4 requires a copy of the Rule 3 notice to be sent with the application, so that time is not wasted in correspondence about the kind of dispute that the conciliator is to be asked to consider. It is thus of importance that the Rule 3 notice should be prepared with proper care in the first place.

Rule 5. The party requesting conciliation shall deliver to the conciliator upon his appointment under Rule 3 or 4 a copy of the Notice prescribed by Rule 3 together with the names and addresses of the parties' representatives.

Rule 6. The conciliator shall start the conciliation as soon as possible after his appointment and shall use his best endeavours to conclude the conciliation as soon as possible and in any event within two months of his appointment unless the parties otherwise agree.

It will usually be in the interest of the party requesting conciliation that the procedure should be as speedy as possible. The rules therefore place upon that party the burden of taking the administrative initiative. Thereafter, the conciliator takes over the conduct of the case, with an express obligation to reach a conclusion as soon as possible.

It follows that, before accepting an appointment, the conciliator-elect should satisfy himself that he has sufficient time available to enable him to conclude the proceedings within the prescribed two month limit. If he cannot embark upon his duties immediately he should reject the appointment.

The two-month period stipulated in Rule 6 is a 'long-stop' provision, since it will be unusual for a conciliation to take this long, unless, of course, the degree of willingness necessary for a successful conciliation is lacking, in which case it may be better to proceed directly to arbitration. However, there may be the odd occasion when, despite the best efforts of all concerned, a conclusion cannot be reached within the two months allowed. In such an event the conciliator must then make such recommendation as he can at that stage or, if both parties still wish to conciliate, he should seek their agreement to an extension of the time limit.

Launching the investigation

Rule 7. Any party may upon receipt of the Notice under Rule 3, or the appointment of the conciliator under Rule 4 (whichever shall be the later) and within such a period as the conciliator may allow send to the conciliator and each other written submissions stating their version of the dispute, difference or other matter together with their views as to the rights and liabilities of the parties arising from it and the financial consequences. Copies of all relevant documents relied on should be attached to any written submission which may be accompanied by a written statement of evidence.

There is in conciliation no express or implied requirement that the parties shall submit their respective cases to the conciliator, since the essence of the procedure is that the conciliator is free to make his own enquiries, normally of the parties themselves but also elsewhere if this will assist him to resolve the dispute quickly. Nevertheless, Rule 7 entitles either party to send 'written submissions stating their version of the dispute' to the conciliator and the other party (or parties) 'within such a period as the conciliator may allow'.

In practice, the conciliator will normally ask each party at the outset whether they wish to send him such statements and, if they do, lay down a programme for their submission. If one party wants to send him a statement but the other party does not, the conciliator should nevertheless still make provision in his programme for a statement in reply, since the party concerned may change its mind when it receives the other's statement. If neither party wishes to send a statement, the conciliator should instead set up a programme of visits or meetings during which he will try to find out by oral means the facts and contentions needed upon which to formulate his recommendation for settlement.

Where a statement or statements are to be submitted, and since they will then provide the basis for the conciliation process, care in their preparation and presentation will be an advantage, particularly during the later stages of the procedure. The best form of submission, therefore, will usually consist of a narrative describing the events in chronological order leading up to the dispute. This should be followed by any argument that the party wishes to put forward in favour of its own position, and finally the statement of the remedy sought, and, if this is a financial sum, the means by which it is calculated. Any supporting documents may conveniently be contained in an appendix with each one numbered and referenced in the submission statement.

Within reason, the more detailed these statements are (while remaining concise and to the point) the easier it should be for the conciliator to grasp the essentials of the dispute in question. Supporting documents must be carefully selected for relevance, and should be kept to the irreducible minimum of key documents, since the conciliator can always request amplification later if he thinks it necessary.

Statements in reply

Bearing in mind the two month time limit under Rule 6, the time allowed for the submission of statements under Rule 7 should be short, and certainly not more than 14 days without a good and overriding reason (although there is no express requirement under the Conciliation Procedure to this effect). In most cases the parties will be able to submit their statements in parallel, and any attempt by one party to withhold its statement until it has had time to consider the other's should be resisted, since a right of reply can always be granted under Rule 8. Such a request may be a delaying tactic, but will more often be attributable to the influence of lawyers familiar with formal pleadings, which are wholly inappropriate to a conciliation.

> Rule 8. With the prior agreement of the conciliator a further period not exceeding 14 days shall be allowed after the period allowed by the conciliator for written submissions under Rule 7 during which any party may send a further written submission to the conciliator and each other replying specifically to points made in any other party's original submission.

The statements of reply should be as short as possible, and confined to matters which have not been dealt with in the parties' original submissions. It is not intended that the reply should be merely a repeat of the original submission. The conciliator is quite capable of understanding the differences where the two submissions contradict each other.

Application

Should the additional points raised in the statements of reply require further documentary evidence, it should accompany the reply submission.

The 14 day period allowed for this process should normally be sufficient, and only when it is necessary to provide an additional document or evidence which will take some days to obtain should an extension of the period be considered.

Further investigation

Rule 9. The conciliator may on his own initiative at any time after his appointment on giving not less than 24 hours' notice to the parties visit and inspect the Site for the subject matter of the dispute. He may generally inform himself in any way he thinks fit of the nature and facts of the dispute, difference or other matter referred to him, including meeting the parties separately.

The conciliator, unlike an arbitrator, may investigate the dispute independently of the parties. He can visit the Site alone or accompanied by one or both of the parties and he may interview the parties separately. Where it appears to the conciliator that one of the parties is in possession of information which he does not wish to disclose to the other, the conciliator is free to discuss such matters with that party and give an undertaking that they will not be disclosed to the other side without agreement. The whole process is one of trying to seek a compromise to the dispute, and if the conciliator understands the fears and/or problems of the parties, he may well be able to suggest a solution which meets the grounds of these fears.

Unlike an arbitrator, the conciliator need not observe the rules of natural justice. Thus neither party has a right to be appraised of all the information which is available to the conciliator, nor is it necessary that each party shall have full knowledge of the case being made against him or the opportunity of replying to every point. Again, while an arbitrator who met one party in the absence of the other would normally be guilty of misconduct, a conciliator may freely do so to assist him in resolving the dispute.

Meeting the parties

Once the conciliator has a firm grasp of the matters in dispute and the background against which they arose, it is often useful (but not a requirement) for him to convene a meeting with both parties to bring everything together, allow the parties to make further oral submissions and generally to 'round off' his investigations. Such a meeting is not a hearing of the kind which would arise in arbitration; rather it is a more or less informal business meeting with the conciliator as chairman.

Rule 10. The conciliator may convene a meeting at which the parties shall be present. He shall give the parties not less than 7 days' notice of such a meeting unless they agree a shorter period. At the meeting the conciliator may take evidence and hear submissions on behalf of any party but shall not be bound by the rules of evidence or by any rules of procedure other than these rules. If it is not possible to conclude the business of the meeting held under this rule on the day or days appointed by him a conciliator may adjourn such meeting to a day to be fixed by him.

The conciliator may convene a meeting at which the parties shall be present . . .

If the conciliation is to be a success, the parties must be represented at the meeting by persons having sufficient authority to take binding decisions on their behalf. One of the subsidiary purposes of the meeting is to encourage the parties to settle their differences themselves, and this cannot happen unless those present have the power to agree a settlement. In any event, experience indicates that the meeting is unlikely to succeed if it is attended only by lawyers or by the junior site staff who became embroiled in the dispute in the first place. Indeed, there is much to be said for the express exclusion of lawyers unless the dispute is mainly about difficult points of law (in which case it will usually be better to proceed direct to arbitration).

It often happens that senior executives of companies do not appreciate the significance of the various arguments until disputes have been referred either to conciliation or arbitration. This meeting is often the first time that they are in a position to appreciate fully the strengths and weaknesses of their own cases. If such persons are not present at the hearing, then when they eventually are required to make a decision, they are likely to do so on the recommendation of those who originally caused the dispute. Under such circumstances, conciliation is likely to fail.

Conciliators should insist that the decision-makers be present and consider carefully the advisability of continuing with the reference if such people are unwilling to attend.

At the meeting the conciliator may take evidence and hear submissions on behalf of any party which shall not be bound by the rules of evidence or by any rules of procedure other than these rules.

As he is not bound by the rules of evidence the conciliator may conduct the taking of evidence in any way he thinks fit. He is not bound to disregard hearsay evidence or evidence which would normally be inadmissible, nor have the parties any right to cross-examination, although the conciliator may allow them to put questions to any witness if he thinks it might be useful, bearing in mind that one of the aims of the meeting should be to promote a compromise. The conciliator may not, however, take evidence on oath.

The witnesses may be examined in the presence of both parties or by themselves in private, and the conciliator may also adjourn the proceedings at any time for separate discussions with the parties. The conciliator may also receive evidence on an undertaking that the information so adduced shall not be passed on to one or other of the parties. However, an undertaking to withhold certain information from both parties should on principle never be given.

To reach a compromise it may be necessary for the conciliator to explore both the strengths and weaknesses of each party's case and, if necessary, explain the situation to them. The conciliator can state confidentially to each party his view on what their success might be if the dispute was taken to arbitration. It would be helpful if at the beginning of the meeting the conciliator tells the parties that the proceedings are conducted on a 'without prejudice' basis, and will not be disclosed to an arbitrator in any subsequent proceedings unless both the parties agree.

The conciliator should also give an undertaking that he will not appear as a witness for either party in any subsequent arbitration.

Application

Rule 11. The conciliator may and shall if requested by all parties seek legal advice or other advice.

Such advice can be sought both prior to the meeting or as a result of it. It may be necessary to adjourn the meeting so that the conciliator can seek advice and then hold a meeting after the advice has become available.

The need for outside advice will most often arise where the dispute turns on some question of law. Outside advice on technical matters is less likely to be needed, since both the parties and the conciliator should normally be able to rely on their own expertise. Nevertheless, where there is a clear difference of view between the parties on some technical aspect of the Contract which is outside the parties' normal expertise it can be useful to refer that particular aspect to an independent expert with an agreement that the parties will abide by his opinion, thereby relieving the conciliator of the need to deal with that part of the problem before him.

Having given the conciliator authority to seek legal or other advice, it is to be implied that the parties have also agreed to bear the cost of such action.

Conciliator's recommendation and opinion

Rule 12. (1) Subject to Rules 6, 7 and 8 and within 21 days of the conclusion of any meeting held pursuant to Rule 10 the conciliator shall prepare his recommendations as to the way in which the matter shall be disposed of and settled between the parties including any recommendation as to any sum of money which should be paid by one party to another.

(2) If the conciliator considers it appropriate so to do he may at the same time or within seven days of giving his recommendations also submit in a separate document his written opinion on the matter or on any part of the matter referred to him. The conciliator's opinion, if given, shall contain such reasons for and comments thereon as in all the circumstances he may deem appropriate.

(3) The conciliator may at any time at his discretion, if he considers it appropriate, or is so requested by the parties, express his preliminary views on any matter referred to him.

. . . the conciliator shall prepare his recommendation . . .

If the parties can be persuaded to reach a settlement of their dispute during the currency of the conciliation, there is no requirement that the conciliator record the terms of that settlement unless the parties expressly request him to do so. However, if no such settlement is reached, the conciliator must bring the proceedings to a close by issuing his own recommendations for settlement.

Under Clause 11.4 of the minor works form the parties must refer the dispute to arbitration within 28 days of receiving the conciliator's recommendation, failing which his recommendation is deemed to have been accepted in settlement of the dispute.

Since the conciliator's recommendation will be final and binding in the absence of a reference to arbitration the recommendation should, like an arbitration award, cover all the matters referred to conciliation, be clear and consistent in all its arguments, and be capable of implementation.

> . . . the conciliator . . . may . . . also submit in a separate document his written opinion on the matter . . .

The conciliator is not bound to give reasons for his recommendations but should do so if so requested by either of the parties. Since conciliation is a persuasive process the reasons may help the 'losing party' to accept the recommendations and the conciliator should give thought to their presentation with this in mind. Alternatively, a reasoned recommendation may often provide a basis for subsequent settlement.

> The conciliator may at any time . . . express his preliminary views on any matter . . .

This rather curious statement seems to have been put in as an afterthought and as a hangover from the arbitration procedure. There is nothing to stop the conciliator opening the meeting between the parties having read their various statements by giving his preliminary views on the dispute. The guiding principle for the conciliator when deciding whether to express an opinion or not should always be whether expressing his views will help resolve the differences.

If a compromise can be achieved at the meeting, then it is advisable for the conciliator to draft with the parties a written statement of the agreement reached. For it is only when expressed in writing that it can finally be certain that agreement has actually been reached.

Publication and fees

> Rule 13. When the conciliator has prepared his recommendation he shall notify the parties in writing and send them an account of his fees and disbursements. Unless otherwise agreed between themselves each party shall be responsible for payment and shall within seven days of receipt of the notice from the conciliator pay an equal share of the amount save that the parties shall be jointly and severally liable to the conciliator for the whole of his account. Upon receipt of payment in full the conciliator shall send his recommendations to all the parties provided that if any party shall fail to make payment due from him any other party may after giving seven days' notice in writing to the defaulting party pay the sum to the conciliator and recover the amount from the defaulting party as a debt due.

The conciliator prepares his recommendations and then advises the parties that his recommendations will be sent to them when they have paid his fees. This procedure is very similar to that adopted by arbitrators when they publish their awards. However, the ICE publishes a form of engagement for a conciliator which binds the parties jointly and severally to pay the conciliator's fees and also indemnifies the conciliator against any actions arising out of the conciliation. When the conciliator has been engaged under the ICE terms then there is no need for the lien on the recommendations which could be sent by agreement to the parties when the conciliator has completed them.

If one of the parties pays the whole of the fees then he is entitled to recover half the sum so paid from the other party.

Application

Conciliator not to be arbitrator

Rule 14. The conciliator shall not be appointed arbitrator in any subsequent arbitration between the parties whether arising out of the dispute, difference or other matters or otherwise arising out of the same Contract unless the parties otherwise agree in writing.

Because the conciliator may have received confidential information separately from each of the parties during the conciliation he will not normally be a suitable person to be arbitrator in any subsequent proceedings on the same matters. Nevertheless, if the parties agree that he should be the arbitrator then there is no difficulty. Problems could arise, however, if the appointment of an arbitrator had to be left to an appointing authority such as the President of the Institution of Civil Engineers. If such an approach was made by the appointing body to the conciliator then the conciliator must refuse to accept the appointment as arbitrator.

Conciliator in subsequent proceedings

There is nothing to stop the conciliator from acting as an expert for either party in any subsequent proceedings but there may be practical grounds for discouraging such appointments. For instance, if in a subsequent arbitration there was a meeting of experts appointed by the two parties it might be hard for a person who had been conciliator not to use confidential information gained during the conciliation when arguing with his opposite number.

There is similar objection to the conciliator acting as a lay advocate for one of the parties in any subsequent proceedings, particularly if the arbitration is to be carried out under the full ICE procedure and not the Short Procedure (Part F) which is stipulated under the minor works form, the reason being that it would be difficult for him when cross examining witnesses not to be mindful of the confidential information that he had obtained during the conciliation.

The best practice would therefore seem to be that (with the exception of a further conciliation) the conciliator should not be involved further in any disputes between the parties arising out of the same contract.

Notices

Rule 15. Any document required by these rules shall be sent to the parties by recorded delivery to the principal place of business or if a company to its registered office. Any documents required by these rules to be sent to the conciliator should be sent by recorded delivery to him at the address which he shall notify to the parties on his appointment.

Again this is a hangover from the arbitration procedure. Since conciliation is a voluntary process, such formalities with the sending of documents would seem to be unnecessary. Provided that the conciliator checks at the meeting that he has received every document that has been sent to him by the parties then how they were sent would not be relevant.

Conclusions

In the minor works form the conciliation procedure takes the place of the reference to the Engineer under Clause 6 of ICE 5. It is a convenient, cheap and speedy way of enabling differences between the parties that have got stuck to be resolved. Very often disputes arise because people have taken up positions from which they find it very difficult to retreat and it needs the intervention of a third party to provide a solution. Conciliation by an expert acceptable to the parties is one of the easiest ways of providing that release.

The ICE Conciliation Procedure is suitable for use with any contract. It is a procedure entered into voluntarily and is suitable for any dispute where there is a will to find a solution.

It is particularly suitable for sorting out the differences between main contractors and their sub-contractors. All that is needed is a simple exchange of letters between the parties agreeing to conciliation and selecting a person to act as conciliator. If the parties do not know of a suitable person then they can always apply to the Arbitration Office of the Institution of Civil Engineers in Great George Street, London for the names of conciliators in the region where the Contract is being executed.

Since conciliation would then not arise under the Contract, the parties must agree the effect of the conciliator's recommendation in the event of a compromise not being reached at the hearing. There is no point in the conciliator making a recommendation unless the parties agree a time period for its acceptance or rejection and whether or not it will be final and binding unless the dispute is taken to arbitration within a fixed period.

Comparison with ICE 5

As there is no provision for conciliation under ICE 5, no comparison is necessary. However conciliation in the minor works form fulfils the function of the reference back to the Engineer for a formal Clause 66(1) decision under ICE 5.

Comparison with ICE 6

While retaining the reference back to the Engineer under ICE 5, Clause 66(1) — re-numbered as Clause 66(3) in ICE 6 — conciliation has been introduced in Clause 66(5) as an optional intermediate stage in the ICE 6 dispute resolution provisions.

Notice of conciliation may be given at any time after the Engineer has given his decision under Clause 66(3) or the time for the giving of that decision under Clause 66(3) has expired, provided that a Notice to Refer to arbitration under Clause 66(6) has not already been served. Save for necessary alterations to cross-referenced clause numbers, the conciliation procedure is the same under ICE 6 as it is under the minor works form, including the provision that the conciliator's recommendation shall be deemed to have been accepted in settlement of the dispute unless a written Notice to Refer under Clause 66(6) is served within one calendar month of its receipt.

In passing, there is a curious lacuna in ICE 6 in that, while Notice to Refer to arbitration must be given within three calendar months of the Engineer's Clause 66(3) decision, after which that decision becomes final and binding, there is no equivalent time limit in Clause 66(5). At first sight it would seem, therefore, that even after the time for reference to arbitration has expired the right to refer to conciliation may endure — in which case could a right to arbitration be revived by that route? However, the true construction of Clause 66 as a whole is probably that the right to arbitration will be lost three months after the Engineer's decision and, as that decision is then final and binding, there will thereafter be nothing left to be decided and thus nothing upon which to conciliate.

Legal comment

Conciliation is, by definition, not a legal procedure, and there is thus nothing on the procedure itself which needs legal comment. Nevertheless, Rule 14 has attracted some adverse comment from lawyers in that in law there is nothing to prevent either party from calling the conciliator as a witness in any subsequent proceedings and that, once on the witness stand and under oath, the conciliator would be bound to reveal any or all of the information he had been given during the conciliation, including that given in confidence. It is argued that this possibility must of necessity make the parties reluctant to co-operate wholeheartedly in the conciliation, which must therefore tend towards an adversarial kind of procedure.

With all due respect to those raising the matter, and bearing in mind that it has never been tested before a court, it is submitted that conciliation is itself clearly a bona fide attempt at settlement or compromise. As such it must carry the benefit of legal privilege, which will entitle the party not wishing certain evidence to be given to claim privilege and thereby exclude it. Given proper legal advice, the parties' legitimate interests should therefore be protected.

Nevertheless, the conciliator should normally decline to appear as a witness in subsequent proceedings unless compelled to do so by a subpoena, in which case the objecting party can apply to have the subpoena withdrawn on the grounds of privilege. If he is so called, it is doubtful whether he could refuse to take the oath (thus avoiding the duty to tell the whole truth), although he should certainly explain his dilemma to the judge or arbitrator before being sworn.

Chapter 20

CONTRACT ADMINISTRATION

Introduction

The prime function of the Engineer and his staff when supervising the construction of works is to safeguard the interests of their client. They must administer all stages in the Contract — preparation of the documents, agreement with the Contractor and the execution of the Works — in such a manner that, so far as is possible, the client gets what he wants at the time he expects to get it, and for the amount that he agreed to pay. None of these objectives can be achieved by instruction alone. They can all be jeopardized by lack of care in the administration of the Contract at any of its stages, or by failure to master detail at the appropriate time.

Administration cannot convert a bad contractor into a good one, but bad administration can ruin an otherwise good contract. This applies irrespective of the size of the project. Even a minor works project can result in a major disaster if not properly administered.

It is not possible to consider all eventualities that may occur in civil engineering works, but there are some basic matters which apply to most contracts and on which detailed consideration will generally be helpful. The aim of this chapter is to assist those responsible for drafting tender and contract documents. The drafting of the contract documents is the second most important task (after the design work) that the Engineer has to perform. Errors in the drafting can be both expensive and disruptive to both the relationship between the Engineer and the Contractor and the progress of the Works. It is therefore necessary that the draughtsman should understand clearly the task he is attempting to perform and should also give a great deal of detailed attention to checking the final draft documents before they are sent to the Contractor.

It should always be borne in mind that interpretation of the resulting contract will not necessarily be in accordance with what the draughtsman intended. Under the legal *contra proferentem* rule ambiguities in exclusion clauses may be interpreted against the party incorporating the clause into the Contract — in this case the Employer.

Contract selection

The factors involved when deciding whether or not to use the minor works form are discussed in Chapter 3. The following matters should, however, be checked before the contract documentation is prepared:

(i) can the Contract be let within two months?

(ii) is the Site available, or will it be available within the two-month period?
(iii) is the design sufficiently advanced to enable all details to be supplied at the starting date?
(iv) is there any specialist design work to be carried out either by the Contractor or by a specialist sub-contractor?
(v) is the Contractor required to let any work to a specified sub-contractor?

If the answers to (i), (ii) and (iii) are 'yes', then it is safe to proceed with the documentation. If 'no', documentation should be postponed until the necessary information is available. If the answers to (iv) and (v) are 'yes' then it will be necessary to define in detail in the Contract the work so required, and this should be borne in mind during the drafting process.

Contract preparation

Administratively, the preparation of the contract documents is probably the most important stage of any contract. Errors in the documents or omissions therefrom will have to be rectified later, normally including an increase in both time and cost. Time spent on checking that all the documents are consistent and that they cover adequately all aspects of the proposed work is seldom wasted.

The criteria for selecting the minor works form are set out in Chapter 3 as is a general description of the contract documents. This chapter will therefore be directed to checklists of matters requiring attention after the minor works form has been selected.

Tender documents

The tender documents will consist of all the documents listed in the Contract Schedule on page 2 of the Contract. In addition, there should be a list of instructions to tenderers. A basic checklist for the contract documents is given in Checklist 1 at the end of this chapter.

1. Covering letter

This letter should contain the formal invitation to the Contractor to submit a tender for the Works. It may then include a general disclaimer stating the Employer does not undertake to accept the lowest or any tender. It may also contain a time period during which the Contractor is expected to hold his tender price open.

For convenience, a list of accompanying documents may also be given. A sample covering letter is given in Appendix E1. A checklist for matter to be included in the covering letter is given in Checklist 2 at the end of this chapter.

Finally, it is convenient to supply an addressed envelope in which the Contractor shall submit his tender. This assists in assuring anonymity of the tenderer and also ensures that tenders are identified when they come into the office and are not opened early by mistake. It is necessary to make sure that the envelope is of sufficient size to contain all the required documents.

2. Instructions to tenderers

The instructions to tenderers are instructions concerning the procedure for the preparation and submission of tenders only and are not intended to form part of the eventual Contract or to have any relevance in its interpretation.

Well-written instructions can greatly assist tenderers in the preparation of their tenders and can lead to cheaper pricing. The more information and guidance that can be given to tenderers the better it will be for the interests of the client. The aim should be to give clear instructions on all matters that the Engineer wishes the tenderer to consider on the manner in which the tender is to be delivered and on the form in which it is to be submitted.

A sample instruction to tenderers is given in Appendix E1, and a list for the matters to be contained in the instructions to tenderers is given in Checklist 3 at the end of this chapter.

3. ICE Conditions of Contract for Minor Works

The conditions may be incorporated by reference in which case the words 'The Conditions of Contract will be *Conditions of Contract, Agreement and Contract Schedule for use in connection with Minor Works of Civil Engineering Construction. First edition (January 1988)*' should be included in the specification.

Normally a copy of the minor works contract should be sent out with the tender documents together with the Contract Schedule and an Appendix to the Conditions of Contract duly completed by the Engineer before dispatch to the tenderers. However, since both these latter documents may need revision for the final Contract, it may be more convenient to submit photocopies of them with the tender document. This is particularly so when a formal agreement is to be entered into, and in due course will be incorporated in the Contract.

A list of items to be sent with the tender documents is to be found in Checklist 4 at the end of this chapter.

4. Special conditions and amendments to the conditions

Normally the standard conditions should not be amended. The conditions have been carefully drafted, and the Contract is designed to be read as a whole. Amendments, however small, can break this continuity and should never be attempted without the guidance of someone who is wholly familiar with the conditions. Special conditions on the other hand are a very useful way of tailoring a set of standard conditions to suit a particular project. However, care must be taken to see that any special conditions do not contradict or affect the standard clauses.

Such matters as restricted access to the Site, services that will be provided by the Employer and any special obligations imposed upon the Contractor are very suitable items to be included as special conditions. So too are any special requirements that the Employer may require in relation to a specific project. Alternatively, such matter may be included in the Specification (if any).

Each special condition should be numbered and the entry in the Contract Schedule (see 8 below) should be: 'Special conditions numbered 1 to *x*.'

5. *Appendix to the Conditions of Contract*

This is an essential document. If it is not completed the Contract is inoperable. The Notes for guidance (Note 13), give advice necessary for the completion of the Appendix, and all items are discussed in detail in Chapter 3. This document should be completed by the Engineer prior to sending out the tender documents.

Short description of the work to be carried out under the Contract. This is important because it defines the limit of the Engineer's powers under the Contract. The Engineer is entitled to give instructions in connection with the Works and although Clause 1.1 defines the Works as including variations ordered by the Engineer, such variations cannot be outside the general scope of the intended Works as defined in the contract documents. It is important therefore that the short description defines clearly the scope of work to be carried out and, where possible, any additions that may be required by the Engineer.

Payment systems available under the Contract. These are set out in Item 2. All that is necessary is for those forms of payment that are not required to be deleted.

'Where a Bill of Quantities or a Schedule of Rates is provided the Method of Measurement used is . . .' The date of the edition that was used for their preparation should also be inserted. Because of the terminology used in methods of measurement certain amendments may be needed in order to ensure compatibility with the minor works form. This can most suitably be done in the preamble to the Bill of Quantities or Schedule of Rates. Suitable amendments for the use of CESMM 2 are given in Chapter 16.

Name of the Engineer. It is suggested in the Notes for guidance (Note 13(1)) that a named person, *not* a firm, should be appointed Engineer under the Contract. The intention is that this person should be in day to day control of the Works and therefore able to carry out the administrative functions necessary. Where possible a name should be supplied at the time of tender.

Starting date. It is advisable that a starting date be included. If not known this item can be left blank.

Period for completion. This can either be filled in by the Engineer or left for the Contractor to complete. If it is to be left for the Contractor to complete, a note to that effect should be included in the instructions to tenderers.

Times for completion of parts. Where a part of the Works is to be handed over before the date of completion for the whole of the Works it must be designated in the Appendix. There is no method by which the Engineer can instruct the Contractor to hand over a part earlier than the date for completion unless such part is defined in the Appendix.

The space provided for defining the parts is limited and if this is insufficient then such parts should be defined in detail elsewhere in the Contract (for instance in the Specification), and reference made in the Appendix to where such definition can be found. The period for completion for each such part is of course measured from the starting date for the Contract as a whole.

There is no provision for liquidated damages in respect of a part.

Liquidated damages. This must be a genuine pre-estimate of the likely damage to be suffered by the Employer if the work is late. The Engineer should make an approximate calculation before completing this item. The estimate should include for such items of damages as

(i) loss of interest on the capital sum invested,
(ii) the cost of overrun of the Engineer's staff,
(iii) the cost of loss of use of the Works including loss of rent, loss of profit from output and additional cost of rent elsewhere for premises not vacated,
(iv) provision for claims for damages for other Contractors affected by late handover,
(v) any damages payable by the Employer as a consequence of late handover, and
(vii) any other form of damage that will arise as a consequence of lateness.

Once the calculation has been made it may be found that the damage is too great for the type of contractor who is likely to undertake the work to be expected to bear. A judgement has therefore to be made as to whether a lesser figure should be included in the Contract which is both acceptable to prospective contractors whilst being of sufficient deterrent effect to satisfy the requirements of the Employer. It has also to be borne in mind that there is a limit to liquidated damages in Item 9 and the figure for liquidated damages per week should therefore not be so great that the limit is reached in too short a time, since its deterrent effect thereafter will be minimal.

As a rough guide it is suggested that liquidated damages should not be more than 1% of the contract value per week.

Limit of liquidated damages. The Notes for guidance suggest that this should not exceed 10% of the contract value. If the advice given in the previous item is followed then the limit for liquidated damages will be reached after ten weeks in relation to a contract not normally exceeding twelve months. This would seem appropriate for most ordinary purposes.

Defects Correction Period. The Notes for guidance suggest that this period should be six months but not exceeding twelve. It should be remembered that the Defects Correction Period starts from the date of practical completion of the whole of the Works and that parts will therefore have a Defects Correction Period in excess of that which is stipulated under Item 10.

Rate and limit of retention. Figures for a rate of retention and its limits are given in Notes for guidance (Note 13(5)).

Minimum amount of interim certificate. It is suggested in the Notes for guidance (Note 13(6)) that the minimum interim certificate should be 10% of the contract value rounded up to the nearest £1000. This figure only applies up to the date of the issue of the Certificate for Practical Completion and thereafter the Contractor is entitled to apply and be certified for any outstanding sums of money due.

It is not absolutely necessary to stipulate a figure and Employers may in practice be willing to pay any sum that is due to the Contractor regardless of size. It is also useful for the Engineer to be supplied on a monthly basis with the value of work actually executed even if this does not reach some stipulated minimum. There are therefore advantages in not providing for a minimum, in which case the items should either be left blank or completed with the words 'not applicable'.

Bank whose base lending rate is used. This should normally be completed by the Contractor and a note to this effect included in the instructions to tenderers. However, if the Employer so wishes he may state the name of an appropriate bank. It is suggested that such a bank should normally be one of the major high street banks.

Insurances. All that is needed is to delete either 'Required' or 'Not required' according to whether or not the Employer is taking out insurance for the care of the Works. If the insurance is taken out by the Employer and 'Required' is deleted then it is necessary that the Contractor's attention is drawn to the conditions attached to the Employer's insurance policy. It is particularly important if the Contractor is to meet any excesses that may be stipulated in the policies that these excesses are clearly defined.

Details of the insurance policies and conditions attached to them should be incorporated either into the Specification or into special conditions if a list of special conditions is to be provided.

Minimum amount of third party insurance. This is an item that should be discussed with the Employer and should reflect the degree of exposure to the Employer should the Contractor cause damage to third parties. Work on highways or airfields or adjacent to railways is likely to carry greater risk of liability than Works in an open field or on a new housing estate.

All that is necessary is that a figure be included in Item 16. The 'Any one accident/Number of accidents unlimited' applies to all contracts and no part of it should be deleted.

6. Drawings

All drawings which are intended to form part of the Contract should be listed, and the list included in the Contract Schedule (see 8 below). The Contractor is entitled to all details on the starting date and therefore all drawings that are to be issued for the purposes of the Contract should be listed here. It may not be necessary to send all drawings out to the tenderers but where only a selection are incorporated with the tender documents reference should be made in the instructions to tenderers (see 2 above) detailing what further drawings are available for inspection at the Engineer's office.

It is essential also that the amendment letter or number for each drawing be stated since it is the exact copy which the Contractor prices which is important for the interpretation of the Contract.

A list for the issue of drawings is to be found in Checklist 7 at the end of this chapter.

7. Pricing schedule

If a copy of the priced schedule is required back with the tender then two copies should be supplied to the tenderer: one for his retention, and one to be returned with the tender.

Daywork will be executed in accordance with the FCEC dayworks schedule unless a schedule for pricing is either included in the Bill of Quantities or a separate schedule is provided with the tender documents for the Contractor to price.

A list for pricing schedules is to be found in Checklist 8 at the end of this chapter.

8. Contract Schedule

This is an extremely important document. It identifies all the documents that will form the Contract. At the time of tender it lists the documents upon which the tenderer's price is to be based. Thereafter it may need amendment between the time of tender and the final Contract by the inclusion of any correspondence that may pass between the parties either before tenders are submitted or in negotiations thereafter.

At the tender stage the Contract Schedule may be incorporated in the tender documents either as a loose leaf photocopy of the appropriate page from the Contract or the actual page in the Contract duly completed. All separately identifiable documents that are sent out with the invitation to tender or are to be included in the subsequent Contract must be listed in it.

A sample completed Contract Schedule is given in Appendix E1.

Contract documents

The contract documents will normally comprise the following

Agreement (Page 1). The Engineer must decide with the Employer whether it is necessary to enter into the formal Agreement contained in the contract documents. It is, of course, perfectly acceptable and legally binding for the Employer to accept the Contractor's tender by a simple letter of acceptance. If, however negotiations take place subsequent to the submission of the Contractor's tender it is advisable that a formal Agreement be entered into since the Contract Schedule will almost certainly need to be amended during those negotiations.

There is no provision on the Agreement for it to be entered into under seal but if that procedure is necessary the Agreement may be altered accordingly. A note to this effect should be included in the instructions to tenderers.

Contractor's tender. This is not of course sent out with the tender documents! It will, however, automatically be incorporated into any contract that arises from the enquiry. The resulting contract will include all the documents that the Contractor returns with his tender, or which are referred to in it unless such documents are expressly excluded. However, if the Contractor chooses to send in his tender on a standard form of his own devising with conditions on the back it would be wise specifically to exclude such conditions in the letter of acceptance, and not to rely simply on the wording of the Contract Schedule.

There is no printed form of tender with the minor works contract, nor legally is one necessary. The Contractor's letter accompanying the returned priced Bills of Quantities, Schedule of Rates or other statement of price is all that is necessary. However, since many Engineers and Employers are used to receiving a form of tender, a suitably worded draft document is given in Appendix E1.

Conditions of Contract and Drawings. For details of the Conditions of Contract, special conditions, the Appendix to the Conditions of Contract, and the Drawings see headings 3–6 respectively above.

Specification (if any). Where a standard specification is used a careful check must be made to ensure that it covers precisely the work which is required to be undertaken. Often standard specifications either omit relevant requirements or have more than one clause which may apply to a particular type of work, and it may not be clear which one defines the particular standard required.

It usually pays to extract from standard specifications the clauses that are to apply, and incorporate just those into the Contract. Such a check will also disclose any gaps that there may be in the standard specification.

The aim should be to define either the document by name or the relevant pages or clauses, if only part of a standard specification is relevant to the Contract. The paragraph in the Contract Schedule is the only statement which incorporates the Specification into the Contract. Care must therefore be taken to ensure that whatever description is given fully defines all the necessary pages of specification that are intended to be incorporated into the Contract.

Pricing Schedule. For details of the Pricing Schedule see heading 7 above.

Following letters. Provision is made for incorporating into the Contract, by means of the Contract Schedule, such relevant correspondence as may have passed between the Engineer and the Contractor during the post-tender, pre-contract period.

If the Contractor provides a tender letter and this is accepted by the Employer then reference should be made to it in the Contract Schedule. Similarly, the Employer's letter of acceptance (if that is to form the Contract between Employer and Contractor) should also be listed amongst the correspondence.

It is not generally good practice to include all the letters from the sometimes lengthy correspondence which may have passed between the Engineer and the Contractor subsequent to the submission of tenders and leading up to agreement. This correspondence is usually both detailed and conflicting and its inclusion can lead to unnecessary argument later in the Contract. A better procedure is to extract from such correspondence the points that are agreed between the parties as applying to the Contract and list those points in a separate document which can then be incorporated into the Contract Schedule as part of the special conditions (see 4 above).

When there have been lengthy negotiations after the submission of a Contractor's tender it is often wise to revise the Contract Schedule and the special conditions. The revised documents should then be sent to the Contractor with a request that he confirm his agreement thereto in writing.

A checklist for completing the Contract Schedule is to be found in Checklist 9 at the end of this chapter.

Tendering procedure

There is no designated procedure for tendering published for use with the minor works contract. However, there is no reason why the procedures set forth in *Guidance on the preparation, submission and consideration of tenders for civil engineering contracts* published by the ICE Conditions of Contract Standing Joint Committee should not be followed. Copies of this publication can be obtained from the Institution of Civil Engineers.

To enable Contracts to be let quickly after the submission of tenders, it is advisable that the contractors tendering should be pre-selected and that open tendering should not be used. A minimum period of four weeks should be allowed for tendering but consideration should be given to increasing this period if the Contractor has to negotiate with a sub-contractor named in the Specification. As a rule, a starting date should be specified in Item 5 of the Appendix to the Conditions of Contract, and only in exceptional circumstances should this be left blank.

The instructions to tenderers, which accompany the tender documents, should draw the tenderers' attention to such matters as

(i) the insurance requirements in accordance with Clause 10,
(ii) the name and telephone number of the person dealing with queries,
(iii) arrangements for inspecting and visiting the Site,
(iv) the procedure to be adopted in presenting and submitting tenders,
(v) the approximate date when the successful tenderer will be informed, and
(vi) a note informing tenderers that qualified tenders will not be considered.

Post-tender phase

Tender adjudication

Tenders need to be checked even though the checks that can be made may be limited, and the action that can be taken following the checks is restricted.

Although tendering procedures should be adhered to where possible, it must be remembered that they are conventions and not legal requirements (unless European Commission regulations or directives dictate otherwise). All post-tender actions by the Engineer and his staff are for the benefit of the Employer and the aim should be to get the best deal possible for the Employer.

All documents submitted by the Contractor must be scrutinized. They all form part of the Contractor's offer, and it is only upon the terms as submitted that they can be accepted without negotiation with the Contractor.

While as a matter of principle qualified tenders are not normally acceptable the Contractor's covering letter may have conditions attached to it which should be checked to see whether those conditions are acceptable to the Employer, and if they are, what effect they may have since they may cost the Employer more money than appears at first sight.

If a programme or method statement is provided with the tender it is necessary to clarify the status of such documents. If the method statement becomes a term of the Contract it can then only be changed by a variation order, which could result in additional cost. For this reason the Contractor should be informed that these additional documents will not form part of the Contract and will not become terms of it. Nevertheless, the Engineer will be deemed to know their content and will normally be under a duty to point out any matters that he is aware are incapable of fulfilment.

All tenders should be checked for arithmetical accuracy. Normally where the product of the rate and the quantity does not equal the amount of the extension it is the rate that is deemed to be correct. Nevertheless, where such an error is appreciable the Contractor's attention should be drawn to the error and he should be asked if he stands by his rate.

The rates should also be checked individually against the rates submitted by other tenderers. Where rates in any tender are appreciably different from those normally to be expected or from those in the other tenders, an appreciation should be made of the consequences of any change that may occur in the quantity. Where it is thought likely that there could be a change in the quantity so as to affect the interest of the Employer adversely, the Engineer should consult with the Contractor with a view to changing the rate for that particular item whilst making adjustments in the other rates to maintain the overall tender price.

If the Contractor is unwilling to make such an adjustment, then the Engineer must assess the likely risk to the Employer and, if this is significantly likely to increase the price of the lowest tender above that of the second, recommend its rejection.

If the Engineer upon scrutiny considers that the Contractor has made a genuine mistake in any of his rates then he should draw the mistake to the Contractor's attention and consider carefully how the matter is to be decided. The simplest approach is to ask the Contractor if, after being told of the mistake, he stands by his tender or wishes to withdraw it.

A checklist for checking tenders is to be found in Checklist 10 at the end of this chapter.

Contract documentation

Any matters of concern revealed by the tender scrutiny should be cleared with the Contractor and their relationship with the Contract clearly understood by both parties. All discussions should be recorded in writing and where these result in changes in the Contractor's tender it will be necessary to incorporate them into the Contract. This is most conveniently done by preparing a list of special conditions covering all the matters so agreed. The list of special conditions can then be incorporated in the Contract by adding it to the Contract Schedule.

Once negotiations have been completed and the tender price and conditions agreed, the preparation of the contract documents can be carried out. All that should be necessary will be the correction of the Bill of Quantities and/or Schedule of Rates and the incorporation of amendments either in the form of letters or a list of special conditions added to the Contract Schedule.

It is most appropriate for the Contract Schedule to be the printed one published in the ICE minor works form especially if the Agreement is to be signed. Once completed, the contract documents should be sent to the Contractor for signing with a note that they must be completed before work is started.

A checklist for completing the contract documents is to be found in Checklist 11 at the end of this chapter.

Post-contract administration

Tender acceptance

Acceptance should be made as soon as possible after the submission of tenders to enable the Contractor to have as much time as possible for the purposes of pre-planning. It is the intention that acceptance should follow within two months of the submission of tenders (see Note 8 in the Notes for guidance).

Since the Works must be fully designed before going out to tender, the Contractor should be in possession of all necessary information to enable the planning to start as soon as notification of acceptance of his tender has been received.

If the starting date is not given in the Contract the Engineer should, if possible, notify the Contractor of the starting date at the time of notification of acceptance of his tender. The starting date must be within 28 days of acceptance of the tender (Clause 4.1). The Engineer should ensure that all the Site is available for the use of the Contractor from the starting date (see Note 9 in the Notes for guidance).

With the notification the Engineer should request

- (i) the name of the Contractor's agent (Clause 3.4),
- (ii) the name of the Statutory Safety Supervisor,
- (iii) the Contractor's programme as specified in Clause 4.3,
- (iv) the name of the Contractor's insurers and proof of the Contractor's insurances (cover note) (Clauses 10.6–7),
- (v) notification of which works the Contractor intends to sub-let (Clause 8.2),
- (vi) a convenient date for an inaugural meeting, and
- (vii) the Contractor to sign the form of Agreement on page 1 of the Contract and return it to the Engineer;

and also provide the Contractor with the following information:

- (i) the name of the Resident Engineer (if any) and the powers delegated to him (Clause 2.2),

(ii) the names of other persons with delegated powers and the limits of such powers (Clause 2.2), and
(iii) any facilities required for other Contractors and the dates from which they are required (Clause 3.9).

The Notes for guidance point out that:

> NG11. . . . it has to be recognized that in a minor works contract the Contractor might have no full-time supervisor on site and the Contractor may ask for instructions to be delivered or sent elsewhere for the attention of his representative. In these circumstances the Contractor has to accept the fact that urgent instructions might in the interests of safety or for some other reason have to be given directly to the Contractor's operatives on site.

The Engineer may not have full-time supervision on site either and it will be necessary for the Contractor's agent and the Resident Engineer to make arrangements to enable them to keep in regular contact for the purposes of inspection, testing and measurement. Such matters should be discussed and arrangements made at the first post-contract meeting.

Finally, the Contract does not state the number of copies of the contract documents that are to be provided to the Contractor. It is suggested that three copies of the contract documents, and of any revised drawings or specifications should be supplied by the Engineer. This is particularly important for drawings since the Contractor is unlikely to have any suitable copying facilities at the Site.

First post-contract meeting

Immediately following the formation of the Contract a meeting should be called between the Engineer and the Contractor to discuss the administration of the Contract. This is particularly important for contracts under the minor works form where the administrative procedures have been reduced to a minimum.

The aim of the first meeting should be to introduce the staff of both sides and to set up a process of communication so that instructions from the Engineer and requests from the Contractor for further information can be dealt with speedily and with confidence.

The success of a contract depends upon the ease with which the staff of the Contractor and the Engineer communicate and where people may be involved in a contract on a part-time basis it is essential that regular contacts be established to prevent avoidable delays. The following matters should be discussed.

1. Engineer's staff

The Contractor should be given the name and telephone number of the Engineer who is to be responsible for the day to day administration of the Contract. If the Employer restricts the authority of the Engineer then such restrictions should be brought to the attention of the Contractor. The Contractor should also be told the names of any resident engineer or inspectors who will be visiting the Works.

2. Contractor's staff

The Contractor should designate a person to be responsible for the Contract and should inform the Engineer of the name, telephone number and address that person.

On minor works contracts the Contractor may not have a full-time representative on site and arrangements must be made for the Engineer to give urgent instructions to the workmen where necessary. The Engineer and the Contractor should discuss how this should be done and how such instructions will be confirmed to the Contractor.

3. Programme

The Contract does not require the Contractor to provide a programme unless requested to do so by the Engineer. The Contractor should normally be asked to do so. If the Engineer knows of any impediments to the Contractor's progress such as non-availability of site or lack of information, this should be brought to the Contractor's attention.

A procedure should be agreed as to how further drawings and information shall be supplied to the Contractor and the method by which Contractors should request further information that they may require.

Progress meetings should be an essential part of the administration of any contract and this is particularly so under the minor works form where communication is reduced to a minimum. A programme for progress meetings should be set up at the initial meeting and it should be agreed who will attend such meetings on behalf of the parties. Normally the Engineer and the Contractor's representative should attend.

A standard agenda for all progress meetings should be agreed as well as the content and time for submission of any report that the Contractor may be required to prepare for these meetings. A suggested draft agenda for progress meetings is to be found at the end of this chapter.

4. Inspections

The minor works form does not require the Contractor to seek the Engineer's permission before covering up any work. Indeed, there are no express requirements for inspection although the Engineer has power to instruct tests to be made under Clause 2.3(b).

There may be particularly significant operations which the Engineer wishes to inspect prior to covering up. Such operations should be identified and arrangements made so that the inspections can be carried out without delaying the Contractor. The Engineer should identify all such operations and a timetable for the inspections can then be agreed either at once or at subsequent progress meetings.

Provided that sufficient notice is given by the Engineer of his requirements, there is no reason why such inspections should have any detrimental effect upon the Contractor's operations. But the Engineer must remember that unless definite provision for inspection is made the Contractor cannot be expected to hold up his operations awaiting the arrival of an inspector. Contractors on the other hand have a duty to comply with all such reasonable requests by the Engineer.

5. Sub-contractors

A procedure for obtaining the consent of the Engineer to sub-contracting parts of the Works must be agreed. The Engineer must not unreasonably withhold consent but that does not relieve the Contractor of the need to supply all the information that the Engineer may reasonably require to enable him to give such consent.

The minimum information that the Engineer should be given is

(i) the extent of the work to be sub-let,
(ii) the name of the sub-contractor,
(iii) any previous experience that the Contractor has of the sub-contractor, and
(iv) the date when the sub-contractor will start and the resources that he will have to make available.

If a standard information sheet can be agreed, it will facilitate the giving of consent during the Contract.

6. Payment

The procedure for the submission of the Contractor's monthly statement should be agreed; both when it is to be presented and to whom. It is also important that a procedure for the notification of claims for additional payments should be agreed.

A suitable draft agenda for the post-contract meeting is given at the end of this chapter.

Regular progress meetings

During the course of the Contract the major administrative tool will be the progress meeting. Since neither party may have permanent representatives upon the Site there will be no opportunity on a daily basis for the senior representatives to meet and to discuss mutual problems. Progress meetings should therefore be established on a regular basis with a standard agenda. The parties can then come to the meeting with the appropriate information so as to minimize the time taken up by such meetings.

There are three major subjects to be discussed at each meeting. The first of such subjects is progress. The Contractor must complete by the completion date and may also be expected to proceed reasonably in accordance with his programme (if any). He is entitled to such information as will enable him to do this. It is important that the Engineer regularly checks that the Contractor is maintaining his progress and that he has all the necessary information that he requires.

At each meeting the progress achieved so far should be discussed and comments made on any impediments to progress that have occurred since the previous meeting. The Contractor should be required to present at each meeting a detailed programme of the period up to the next meeting. This will enable the Engineer to arrange for any inspection to be carried out or supervision to be provided for any special operations that will be taking place.

It should be clearly understood between the Contractor and Engineer that if the Contractor's progress deviates materially from his detailed programme he will inform the Engineer accordingly. There is no benefit to the Contractor if the Engineer's inspectors turn up either far too early for an operation, or after it has taken place.

The second subject for discussion is that of inspections. Inspections fall into two broad categories:

(i) those that are essential and for which the Employer is prepared to pay for delay should the inspectors be late, and
(ii) those less essential which the Engineer requires but for which he is unwilling to delay the Contractor.

For the former the Contractor should be given instructions to inform the Engineer when he thinks the Works will be available for inspection and that will under no circumstances proceed with the work until the inspection has been carried out. The results of such inspections and any tests that are made should be discussed at the following progress meeting.

The third subject is payment. Normally measurement does not produce many problems for either the Engineer or the Contractor, and such matters can be dealt with speedily. However, this may not be true when agreeing payment for events which cause the Contractor delay or disrupt his operations.

The Contractor should be encouraged to submit claims for additional payment speedily and time should be devoted at the progress meeting to discussing the consequential effects of any instructions that have been issued by the Engineer. Since the payment clause and the extension of time clauses under the minor works form are closely related the Contractor should also submit his claims for extensions of time for matters that have occurred since the previous progress meeting.

The minutes of the progress meetings will become a record of the progress achieved on the Contract and regular discussion of factors causing delay will identify, whilst memories are still good, the resources that may have been affected. This will facilitate the later settlement of any cost claims, and give support to any answers that may have to be given in reply to questions from the Employer or his auditors.

Since the minutes of the meeting will record the programme for the next period and any instructions that may have been given at the meeting to the Contractor with regard to such matters as record-keeping and inspections, it is essential that the minutes should be circulated as quickly as possible. In general the minutes should be circulated within three days of the meeting. It should also be an established principle that the minutes of the meeting are accepted unless objected to in writing within a further three days after receipt.

A regular circulation for the minutes should be established at the beginning of the Contract and this should not be changed merely because different people may be at any particular meeting. Varying the initial circulation always leads to both confusion and delay. If additional parties or guests are present then it will be for the party inviting them to send them a copy of the minutes, if he sees fit and after this has been agreed between both sides.

Final account

All measurements that are made on site should be agreed at the time by the Engineer and by the Contractor. As work is completed its measurement should also be agreed so that agreed quantities for all contract work are available at the same time as the Works are completed. It will be found both easier and less time consuming to agree measurement as work proceeds and not leave it until the Works have been completed.

Checklists

1. *Tender documents*

The tender documents to be sent to the Contractor are

(i) covering letter of invitation,
(ii) instructions to tenderers,
(iii) the minor works contract (printed form),
(iv) special conditions of contract (including amendments to the minor works form),
(v) the appendix to the form of tender (duly completed),
(vi) the Drawings,
(vii) the priced Bill of Quantities and/or Schedule of Rates and/or Daywork Schedule, and
(viii) the Contract Schedule (partially completed)

2. *Covering letter*

The covering letter may contain all or any of the following:

(i) invitation to submit a tender;
(ii) disclaimer (no guarantee to award any contract);
(iii) requirement to enter into a formal contract agreement;
(iv) accompanying documents:

(a) Conditions of Contract,
(b) Specification,
(c) Drawings,
(d) ground investigation information,

and separate documents:

(a) instructions to tenderers,
(b) special conditions,
(c) Bills of Quantities/Schedule of Rates/Daywork Schedule.

3. *Instructions to tenderers*

Instructions to tenderers will usually contain the following:

(i) the documents to be submitted with the tender;
(ii) the venue, time and date for the return of tenders;
(iii) the procedure for delivering and presenting tenders;
(iv) the name and telephone number of the person dealing with enquiries;

(v) the requirements representing rates and prices in the Bill of Quantities/Schedule of Rates:

(a) price in words (for lump sum contracts),
(b) price in figures,
(c) items not priced or marked 'nil' or 'included' will all be deemed to be covered by the other rates and prices,
(d) no alteration to be made to the quantities or item descriptions in the Bill of Quantities or to any of the other tender documents unless directed by the Engineer;

(vi) acceptability of

(a) qualified tenders,
(b) alternative tenders;

(vii) arrangements for visiting the Site;
(viii) approximate starting date (if not given in the Appendix);
(ix) how errors in the bills or schedules will be dealt with;
(x) items in the Appendix that the Contractor is required to fill in (if not stated by the Engineer):

(a) Item 6 (Period for completion),
(b) Item 14 (Bank whose base lending rate is to be used);

(xi) insurance requirements/procedures; and
(xii) borehole cores and other information available for inspection.

There should be added to this list any points of particular interest which should be brought to the attention of a future contractor of which the Engineer is aware. For instance, if further information is available at the Engineer's office, or the Contractor is required to visit any statutory undertaker, or other sources of information which are available, this should be stated.

It is essential that tenderers be supplied with as much information concerning the Works as is available to the Engineer so that prices when submitted will be based upon the widest possible knowledge of the factors affecting the Works.

4. Minor works contract

The following documents must be sent to each tenderer:

(i) for tendering, either

(a) a full copy of the minor works form, or
(b) photocopies of the Contract Schedule, and Appendix to the Conditions of Contract and then incorporate the Conditions of Contract for Minor Works by reference;

(ii) the Contract Schedule completed; and
(iii) the Appendix to the Conditions of Contract completed.

5. Special conditions

Special conditions may include:

(i) variation in price provision;
(ii) any labour, materials or services that can, or will be supplied by the Employer;
(iii) any time restrictions where work either must or must not be carried out;
(iv) conditions as to access;
(v) special provisions as to noise, nuisance or pollution;
(vi) special permissions or permits that may be required;
(vii) conditions as to insurance, including

 (a) policies required,
 (b) excesses,
 (c) production of proof, and
 (d) special insurers;

(viii) arrangements for inspection of completed work;
(ix) special arrangements concerning payment;
(x) arrangements concerning statutory undertakers, services contractors, or other contractors employed by the Employer; and
(xi) arrangements for handing over.

6. *Appendix to the Conditions of Contract*

The draughtsman should ensure that in the Appendix to the Conditions of Contract

(i) all items are completed, and
(ii) the Appendix has been included in the tender documents.

7. *Drawings*

The following is a list of matters should be considered when issuing the Drawings:

(i) are all the Drawings listed?
(ii) are they the latest amendments?
(iii) does the instruction to tenderers state

 (a) the numbers of drawings sent with the tender documents?
 (b) that other drawings are available for inspection?

(iv) schedules to be included:

 (a) reinforcement,
 (b) drainage,
 (c) pre-cast or prefabricated units, and
 (d) other schedules; and

(v) any standard details that are required to be included.

8. *Priced schedules*

The following matters should be considered in connection with any schedules to be priced:

(i) do the bills and/or schedules comply with the stated method of measurement?
(ii) does the preamble to the bill list all deviations from the standard method of measurement?
(iii) is there provision for the Contractor to price dayworks?
(iv) do the instructions to tenderers cover items that are either unpriced or marked 'included'?

9. *Contract Schedule*
The following matters should be considered when drafting the Contract Schedule:

(i) whether the Agreement is required, if this should be stated;
(ii) conditions of contract — are there any amendments?
(iii) are the drawing numbers filled in or reference made to a separate list?
(iv) Specification:

  (a) if standard give reference,
  (b) if special give title,
  (c) if an extract from a standard specification, state pages;

(v) is the form of tender required, and if so, is it included?

10. *Tender check*
The following matters should be considered when checking tenders:

(i) is the Bill of Quantities and/or Schedule of Rates arithmetically correct?
(ii) are there any anomalous rates?
(iii) is the tender accompanied by a letter, and if so, are its contents acceptable?
(iv) is there a programme, and if so

  (a) can the Engineer supply information to meet it?
  (b) does it comply with the requirements of the Contract?
  (c) are there any matters on it with which the Employer could not comply?
  (d) is it necessary to reject it?

(v) is there a method statement with the tender, and if so,

  (a) are the methods acceptable?
  (b) do they comply with the requirements of the specification?
  (c) what are the consequences if the Engineer requires the methods to be changed?

(vi) is all information available to prepare the contract documents?

11. *Contract documents*

(i) has the Agreement been completed on behalf of the Employer?

(ii) The Contract:

(a) has the Contract Schedule been completed?
(b) is the Contractor's tender to form part of the Contract? If not, delete the Contractor's tender from the Contract Schedule.

(iii) Conditions of Contract are there any special conditions to be incorporated? If so, reference should be made to the additional conditions.
(iv) The Appendix to the Conditions of Contract:

(a) is the starting date still applicable?
(b) any change to the periods for completion?
(c) does Item 12 (Limit of retention) need filling-in?

(v) Drawings:

(a) are there any amendments needed to the Drawings before their inclusion in the Contract?
(b) are there any new drawings since the tender, and if so, have they been included?

(vi) The Specification: are there any amendments?
(vii) The priced documents:

(a) have they been amended?
(b) does any reference need to be made on the Contract Schedule?

(viii) Additional documents: is there any correspondence that needs to be incorporated in the Contract from

(a) pre-tender letters, or
(b) post-tender negotiations?

Agendas

Draft agenda for post-contract meeting

(i) Engineer's staff
(ii) Contractor's staff
(iii) Programme:

(a) provision of the programme
(b) progress meetings
(c) supply of information
(d) progress reports

(iv) Inspections:

(a) operations requiring inspection
(b) notification of availability

(v) Sub-contractors: procedure for consent
(vi) Measurement:

(a) dates of monthly statement
(b) procedures for additional payments

Draft agenda for progress meetings

(i) Progress achieved to date
(ii) Detailed programme for the coming period:

(a) inspections required
(b) tests to be taken

(iii) Advance look at the programme for the following period:

(a) any proposed amendments
(b) matters of particular concern

(iv) Details of information required or outstanding
(v) Financial matters:

(a) current interim certificate
(b) rates for varied work
(c) extension of time applications
(d) claims for delay or disruption or unforeseen circumstances
(e) counter claims by the Employer

(vi) Any other business
(vii) Next meeting:

(a) date and location
(b) any special topics to be raised
(c) any special person required to be present

Appendix A

ICE CONDITIONS OF CONTRACT FOR MINOR WORKS

including Notes for guidance

The Institution of Civil Engineers

Conditions of Contract, Agreement and Contract Schedule

for use in connection with

Minor Works

of

Civil Engineering Construction

First edition (January 1988)

The Institution of Civil Engineers Conditions of Contract for Minor Works

Agreement

THIS AGREEMENT is made the .... day of .... 19 ....

between ....

of (or whose registered office is at) ....

....

(hereinafter called the 'Employer') of the one part

and ....

of (or whose registered office is at) ....

....

(hereinafter called the 'Contractor') of the other part

WHEREAS the Employer wishes to have carried out the following ....

....

and has accepted a Tender by the Contractor for the same

NOW IT IS HEREBY AGREED AS FOLLOWS:

Article 1 The Contractor will subject to the Conditions of Contract perform and complete the Works.

Article 2 The Employer will pay the Contractor such sum or sums as shall become payable under the Contract and in accordance with the Conditions of Contract.

Article 3 The documents listed in the Contract Schedule form part of this Agreement.

AS WITNESS the hands of the parties hereto:

Signed for and on behalf of the Employer ....

in the presence of ....
(Witness)

Signed for and on behalf of the Contractor ....

in the presence of ....
(Witness)

Appendix A

The Institution of Civil Engineers Conditions of Contract for Minor Works

The Contract Schedule

(List of documents forming part of the Contract)

The Agreement (if any)

The Contractor's Tender (excluding any general or printed terms contained or referred to therein unless expressly agreed in writing to be incorporated in the Contract)

The Conditions of Contract

The Appendix to the Conditions of Contract

The Drawings. Reference numbers ....

....

....

The Specification. Reference ....

The priced Bill of Quantities*

The Schedule of Rates*

The Daywork Schedules*

The following letters*

from .... to .... dated ....

from .... to .... dated ....

from .... to .... dated ....

from .... to .... dated ....

*Delete if not applicable

The Institution of Civil Engineers Conditions of Contract for Minor Works

Definitions

1. Definitions

1.1. 'Works' means all work necessary for the completion of the Contract including any variations ordered by the Engineer.

1.2. 'Contract' means the Agreement if any together with these Conditions of Contract the Appendix and the other items listed in the Contract Schedule.

1.3. 'Cost' (except for 'cost plus fee' contracts, see Appendix) includes overhead costs whether on or off the Site of the Works but not profit.

1.4. 'Site' means the lands and other places on under in or through which the Works are to be executed and any other lands or places provided by the Employer for the purposes of the Contract.

1.5. 'Excepted Risks' are riot war invasion act of foreign enemies hostilities (whether war be declared or not) civil war rebellion revolution insurrection or military or usurped power ionizing radiations or contamination by radioactivity from any nuclear fuel or from any nuclear waste from the combustion of nuclear fuel radioactive toxic explosive or other hazardous properties of any explosive nuclear assembly or nuclear component thereof pressure waves caused by aircraft or other aerial devices travelling at sonic or supersonic speeds or a cause due to the use or occupation by the Employer his agents servants or other contractors (not being employed by the Contractor) of any part of the Permanent Works or to fault defect or error or omission in the design of the Works (other than a design provided by the Contractor pursuant to his obligations under the Contract).

2. Engineer

Engineer to be a named individual

2.1. The Employer shall appoint and notify to the Contractor in writing a named individual to act as Engineer. If at any time the Engineer is unable to continue the duties required by the Contract the Employer shall forthwith appoint a replacement and shall so notify the Contractor in writing.

2.2. The Engineer may appoint a named Resident Engineer and/or other suitably experienced person to watch and inspect the Works and the Engineer may delegate to such person in writing any of the powers of the Engineer herein provided that prior notice in writing is given to the Contractor.

Engineer's power to give instructions

2.3. The Engineer shall have power to give instructions for:-

(a) any variation to the Works including any addition thereto or omission therefrom;
(b) carrying out any test or investigation;
(c) the suspension of the Works or any part of the Works in accordance with Clause 2.6;
(d) any change in the intended sequence of the Works;
(e) measures necessary to overcome or deal with any obstruction or condition falling within Clause 3.8;
(f) the removal and/or re-execution of any work or materials not in accordance with the Contract;
(g) the elucidation or explanation of any matter to enable the Contractor to meet his obligations under the Contract;
(h) the exclusion from the Site of any person employed thereon which power shall not be exercised unreasonably.

2.4. The Engineer or Resident Engineer and/or other suitably experienced person who exercises any delegated power shall upon the written request of the Contractor specify in writing under which of the foregoing powers any instruction is given. If the Contractor shall be dissatisfied with any such instruction he shall be entitled to refer the matter to the Engineer for his decision.

Dayworks

2.5. The Engineer may order in writing that any work shall be executed on a daywork basis. Subject to the production of proper records the Contractor shall then be entitled to be paid in accordance with a Daywork Schedule included in the Contract or otherwise in accordance with the 'Schedules of Dayworks carried out incidental to Contract Work' issued by the Federation of Civil Engineering Contractors and current at the date the work is carried out.

Engineer may suspend the progress of the Works

2.6. (1) The Engineer may order the suspension of the progress of the Works or any part thereof:-

(a) for the proper execution of the work;
(b) for the safety of the Works or any part thereof;
(c) by reason of weather conditions;

and in such event may issue such instructions as may in his opinion be necessary to protect and secure the Works during the period of suspension.

(2) If permission to resume work is not given by the Engineer within a period of 60 days from the date of the written Order of Suspension then the Contractor may serve a written notice on the Engineer requiring permission to proceed with the Works within 14 days from the receipt of such notice. Subject to the Contractor not being in default under the Contract the Engineer shall grant such permission and if such permission is not granted the Contractor may by a further written notice served on the Engineer elect to treat the suspension where it affects a part of the Works as an omission under Clause 2.3(a) or where the whole of the Works is suspended as an abandonment of the Contract by the Employer.

Parties bound by Engineer's instructions

2.7. Each party shall be bound by and give effect to every instruction or decision of the Engineer unless and until either:-

(a) it is altered or amended by an agreed settlement following a reference under Clause 11.3 and neither party gives notice of dissatisfaction therewith

or

(b) it is altered or amended by a decision of an arbitrator under Clause 11.4 or 11.5.

3. General obligations

Contractor to perform and complete the Works

3.1. The Contractor shall perform and complete the Works and shall (subject to any provision in the Contract) provide all supervision labour materials plant transport and temporary works which may be necessary therefor.

Contractor responsible for care of the Works

3.2. (1) The Contractor shall take full responsibility for the care of the Works from commencement until 14 days after the Engineer issues a Certificate of Practical Completion for the whole of the Works pursuant to Clause 4.5.

(2) If the Engineer issues a Certificate of Completion in respect of any part of the Works before completion of the whole of the Works the Contractor shall cease to be responsible for the care of that part of the Works 14 days thereafter and the responsibility for its care shall then pass to the Employer.

(3) The Contractor shall take full responsibility for the care of any outstanding work which he has undertaken to finish during the Defects Correction Period until such outstanding work is complete.

Contractor to repair and make good at his own expense

3.3. (1) In case any damage loss or injury from any cause whatsoever (save and except the Excepted Risks) shall happen to the Works or any part thereof while the Contractor is responsible for their care the Contractor shall at his own cost repair and make good the same so that at completion the Works shall be in good order and condition and conform in every respect with the requirements of the Contract and the Engineer's instructions.

(2) To the extent that any damage loss or injury arises from any of the Excepted Risks the Contractor shall if required by the Engineer repair and make good the same at the expense of the Employer.

(3) The Contractor shall also be liable for any damage to the Works occasioned by him in the course of any operations carried out by him for the purpose of completing outstanding work or complying with his obligations under Clauses 4.7 and 5.2.

Contractor's authorized representative

3.4. The Contractor shall notify the Engineer of the person duly authorized to receive instructions on behalf of the Contractor.

Setting out and safety of site operations

3.5. The Contractor shall take full responsibility for the setting out of the Works and for the adequacy stability and safety of his site operations and methods of construction.

Engineer to provide necessary information

3.6. Subject to Clause 3.5 the Engineer shall be responsible for the provision of any necessary instructions drawings or other information.

Contractor's responsibility for design

3.7. The Contractor shall not be responsible for the design of the Works except where expressly stated in the Contract. The Contractor shall be responsible for the design of any temporary works other than temporary works designed by the Engineer.

Adverse physical conditions and artificial obstructions — delay and extra cost

3.8. If during the execution of the Works the Contractor shall encounter any artificial obstruction or physical condition (other than a weather condition or condition due to weather) which obstruction or condition could not in his opinion reasonably have been foreseen by an experienced contractor the Contractor shall as early as practicable give written notice thereof to the Engineer. If in the opinion of the Engineer such obstruction or condition could not reasonably have been foreseen by an experienced contractor then the Engineer shall certify and the Employer shall pay a fair and reasonable sum to cover the cost of performing any additional work or using any additional plant or equipment together with a reasonable percentage addition in respect of profit as a result of:-

(a) complying with any instructions which the Engineer may issue

and/or

(b) taking proper and reasonable measures to overcome or deal with the obstruction or condition in the absence of instructions from the Engineer

together with such sum as shall be agreed as the additional cost to the Contractor of the delay or disruption arising therefrom. Failing agreement of such sums the Engineer shall determine the fair and reasonable sum to be paid.

Facilities for other contractors

3.9. The Contractor shall in accordance with the requirements of the Engineer afford reasonable facilities for any other contractor employed by the Employer and for any other properly authorized authority employed on the Site.

4. Starting and completion

Starting date to be notified in writing

4.1. The starting date shall be the date specified in the Appendix or if no date is specified a date to be notified by the Engineer in writing being within a reasonable time and in any event within 28 days after the date of acceptance of the Tender. The Contractor shall begin the Works at or as soon as reasonably possible after the starting date.

Time for completion

4.2. The period or periods for completion shall be as stated in the Appendix or such extended time as may be granted under Clause 4.4 and shall commence on the starting date.

Contractor's programme

4.3. The Contractor shall within 14 days after the starting date if so required provide a programme of his intended activities. The Contractor shall at all times proceed with the Works with due expedition and reasonably in accordance with his programme or any modification thereof which he may provide or which the Engineer may request.

Extension of time for completion

4.4. If the progress of the Works or any part thereof shall be delayed for any of the following reasons:-

(a) an instruction given under Clause 2.3 (a) (c) or (d);
(b) an instruction given under Clause 2.3 (b) where the test or investigation fails to disclose non-compliance with the Contract;
(c) encountering an obstruction or condition falling within Clause 3.8 and/or an instruction given under Clause 2.3 (e);
(d) delay in receipt by the Contractor of necessary instructions drawings or other information;
(e) failure by the Employer to give adequate access to the Works or possession of land required to perform the Works;
(f) delay in receipt by the Contractor of materials to be provided by the Employer under the Contract;
(g) exceptional adverse weather;
(h) other special circumstances of any kind whatsoever outside the control of the Contractor

then provided that the Contractor has taken all reasonable steps to avoid or minimize the delay the Engineer shall upon a written request by the Contractor promptly by notice in writing grant such extension of the period for completion of the whole or part of the Works as may in his opinion be reasonable. The extended period or periods for completion shall be subject to regular review provided that no such review shall result in a decrease in any extension of time already granted by the Engineer.

Certificate of Completion of Works or parts of Works

4.5. (1) Practical completion of the whole of the Works shall occur when the Works reach a state when notwithstanding any defect or outstanding items therein they are taken or are fit to be taken into use or possession by the Employer.

(2) Similarly practical completion of part of the Works may also occur but only if it is fit for such part to be taken into use or possession independently of the remainder.

(3) The Engineer shall upon the Contractor's request promptly certify in writing the date upon which the Works or any part thereof has reached practical completion or otherwise advise the Contractor in writing of the work necessary to achieve such completion.

Liquidated damages

4.6. If by the end of the period or extended period or periods for completion the Works have not reached practical completion the Contractor shall be liable to the Employer in the sum stated in the Appendix as liquidated damages for every week (or pro rata for part of a week) during which the Works so remain uncompleted up to the limit stated in the Appendix. Similarly where part or parts of the Works so remain uncompleted the Contractor shall be liable to the Employer in the sum stated in the Appendix reduced in proportion to the value of those parts which have been certified as complete provided that the said limit shall not be reduced.

Rectification of defects

4.7. The Contractor shall rectify any defects and complete any outstanding items in the Works or any part thereof which reach practical completion promptly thereafter or in such manner and/or time as may be agreed or otherwise accepted by the Engineer. The Contractor shall maintain any parts which reach practical completion in the condition required by the Contract until practical completion of the whole of the Works fair wear and tear excepted.

5. Defects

Definition of Defects Correction Period

5.1. 'Defects Correction Period' means the period stated in the Appendix which period shall run from the date certified as practical completion of the whole of the Works or the last period thereof.

Cost of remedying defects

5.2. If any defects appear in the Works during the Defects Correction Period which are due to the use of materials or workmanship not in accordance with the Contract the Engineer shall give written notice thereof and the Contractor shall make good the same at his own cost.

Remedy for Contractor's failure to correct defects

5.3. If any such defects are not corrected within a reasonable time by the Contractor the Employer may after giving 14 days' written notice to the Contractor employ others to correct the same and the cost thereof shall be payable by the Contractor to the Employer.

Engineer to certify completion

5.4. Upon the expiry of the Defects Correction Period and when any outstanding work notified to the Contractor under Clause 5.2 has been made good the Engineer shall upon the written request of the Contractor certify the date on which the Contractor completed his obligations under the Contract to the Engineer's satisfaction.

Unfulfilled obligations

5.5. Nothing in this Clause shall affect the rights of either party in respect of defects appearing after the Defects Correction Period.

6. Additional payments

Engineer to determine additional sums and deductions

6.1. If the Contractor carries out additional works or incurs additional cost including any cost arising from delay or disruption to the progress of the Works as a result of any of the matters referred to in paragraphs (a) (b) (d) (e) or (f) of Clause 4.4 the Engineer shall certify and the Employer shall pay to the Contractor such additional sum as the Engineer after consultation with the Contractor considers fair and reasonable. Likewise the Engineer shall determine a fair and reasonable deduction to be made in respect of any omission of work.

Valuation of additional work

6.2. In determining a fair and reasonable sum under Clause 6.1 for additional work the Engineer shall have regard to the prices contained in the Contract.

7. Payment

Valuation of the Works

7.1. The Works shall be valued as provided for in the Contract.

Monthly statements

7.2. The Contractor shall submit to the Engineer at intervals of not less than one month a statement showing the estimated value of the Works executed up to the end of that period together with a list of any goods or materials delivered to the Site and their value and any other items which the Contractor considers should be included in an interim certificate.

Interim payments

7.3. Within 28 days of the delivery of such statement the Engineer shall certify and the Employer shall pay to the Contractor such sum as the Engineer considers properly due less retention at the rate of and up to the limit set out in the Appendix. Until practical completion of the whole of the Works the Engineer shall not be required to certify any payment less than the sum stated in the Appendix as the minimum amount of interim certificate. The Engineer may by any certificate delete correct or modify any sum previously certified by him.

Payment of retention money

7.4. One half of the retention money shall be certified by the Engineer and paid to the Contractor within 14 days after the date on which the Engineer issues a Certificate of Practical Completion of the whole of the Works.

7.5. The remainder of the retention money shall be paid to the Contractor within 14 days after the issue of the Engineer's certificate under Clause 5.4.

Contractor to submit final account

7.6. Within 28 days after the issue of the Engineer's certificate under Clause 5.4 the Contractor shall submit a final account to the Engineer together with any documentation reasonably required to enable the Engineer to ascertain the final contract value. Within 42 days after the receipt of this information the Engineer shall issue the final certificate. The Employer shall pay to the Contractor the amount due thereon within 14 days of the issue of the final certificate.

7.7. The final certificate shall save in the case of fraud or dishonesty relating to or affecting any matter dealt with in the certificate be conclusive evidence as to the sum due to the Contractor under or arising out of the Contract (subject only to Clause 7.9) unless either party has within 28 days after the issue of the final certificate given notice under Clause 11.2.

Interest on overdue payments

7.8. In the event of failure by the Engineer to certify or the Employer to make payment in accordance with the Contract the Employer shall pay to the Contractor interest on the amount which should have been certified or paid on a daily basis at a rate equivalent to 2% per annum above the base lending rate of the bank specified in the Appendix.

Value-added tax

7.9. In addition to sums due otherwise to the Contractor under the Contract and notwithstanding any time for payment stipulated in the Contract the Employer shall pay to the Contractor any value-added tax properly chargeable by the Commissioners of Customs and Excise on the supply to the Employer of any goods and/or services by the Contractor under the Contract.

8. Assignment and sub-letting

Assignment

8.1. Neither the Employer nor the Contractor shall assign the Contract or any part thereof or any benefit or interest therein or thereunder without the written consent of the other party.

No sub-letting without Engineer's consent

8.2. The Contractor shall not sub-let the whole of the Works. The Contractor shall not sub-let any part of the Works without the consent of the Engineer which consent shall not be unreasonably withheld.

Contractor responsible for sub-contractors

8.3. The Contractor shall be responsible for any acts defaults or neglects of any sub-contractor his agents servants or workmen in the execution of the Works or any part thereof as if they were the acts defaults or neglects of the Contractor.

9. Statutory obligations

Contractor to comply with statutory requirements

9.1. The Contractor shall subject to Clause 9.3 comply with and give all notices required by any statute statutory instrument rule or order or any regulation or by-law applicable to the construction of the Works (hereinafter called 'the statutory requirements') and shall pay all fees and charges which are payable in respect thereof.

Employer to obtain consents

9.2. The Employer shall be responsible for obtaining in due time any consent approval licence or permission but only to the extent that the same may be necessary for the Works in their permanent form.

Contractor's exemption from liability to comply with statutes

9.3. The Contractor shall not be liable for any failure to comply with the statutory requirements where and to the extent that such failure results from the Contractor having carried out the Works in accordance with the Contract or with any instruction of the Engineer.

10. Liabilities and insurance

Insurance of the Works

10.1. (1) If so stated in the Appendix the Contractor shall maintain insurance in the joint names of the Employer and the Contractor in respect of the Permanent Works and the Temporary Works (including for the purpose of this clause any unfixed materials or other things delivered to the Site for incorporation therein) to their full value and the constructional plant to its full value against all loss or damage from whatever cause arising (other than the Excepted Risks) for which he is responsible under the terms of the Contract.

(2) Such insurance shall be effected in such a manner that the Employer and the Contractor are covered for the period stipulated in Clause 3.2 and are also covered for loss or damage arising during the Defects Correction Period from such cause occurring prior to the commencement of the Defects Correction Period and for any loss or damage occasioned by the Contractor in the course of any operation carried out by him for the purpose of complying with his obligations under Clauses 4.7 and 5.2.

(3) The Contractor shall not be liable to insure against the necessity for the repair or reconstruction of any work constructed with materials or workmanship not in accordance with the requirements of the Contract.

Contractor to insure against damage to persons and property

10.2. The Contractor shall indemnify and keep the Employer indemnified against all losses and claims for injury or damage to any person or property whatsoever (save for the matters for which the Contractor is responsible under Clause 3.2) which may arise out of or in consequence of the Works and against all claims demands proceedings damages costs charges and expenses whatsoever in respect thereof or in relation thereto subject to Clauses 10.3 and 10.4.

10.3. The liability of the Contractor to indemnify the Employer under Clause 10.2 shall be reduced proportionately to the extent that the act or neglect of the Engineer or the Employer his servants or agents or other contractors not employed by the Contractor may have contributed to the said loss injury or damage.

10.4. The Contractor shall not be liable for or in respect of or to indemnify the Employer against any compensation or damage for or with respect to:-

(a) damage to crops being on the Site (save in so far as possession has not been given to the Contractor);
(b) the use or occupation of land (which has been provided by the Employer) by the Works or any part thereof or for the purpose of constructing completing and maintaining the Works (including consequent losses of crops) or interference whether temporary or

permanent with any right of way light air or water or other easement or quasi easement which are the unavoidable result of the construction of the Works in accordance with the Contract;

(c) the right of the Employer to construct the Works or any part thereof on over under in or through any land;

Contractor not liable for unavoidable damage

(d) damage which is the unavoidable result of the construction of the Works in accordance with the Contract;

(e) injuries or damage to persons or property resulting from any act or neglect or breach of statutory duty done or committed by the Engineer or the Employer his agents servants or other contractors (not being employed by the Contractor) or for or in respect of any claims demands proceedings damages costs charges and expenses in respect thereof or in relation thereto.

Employer to indemnify Contractor

10.5. The Employer will save harmless and indemnify the Contractor from and against all claims demands proceedings damages costs charges and expenses in respect of the matters referred to in Clause 10.4. Provided always that the Employer's liability to indemnify the Contractor under paragraph (e) of Clause 10.4 shall be reduced proportionately to the extent that the act or neglect of the Contractor or his subcontractors servants or agents may have contributed to the said injury or damage.

Employer to approve insurance

10.6. The Contractor shall throughout the execution of the Works maintain insurance against damage loss or injury for which he is liable under Clause 10.2 subject to the exceptions provided by Clauses 10.3 and 10.4. Such insurance shall be effected with an insurer and in terms approved by the Employer (which approval shall not be unreasonably withheld) for at least the amount stated in the Appendix. The terms of such insurance shall include a provision whereby in the event of any claim in respect of which the Contractor would be entitled to receive indemnity under the policy being brought or made against the Employer the insurer will indemnify the Employer against any such claims and any costs charges and expenses in respect thereof.

Contractor to produce policies of insurance

10.7. The Contractor shall comply with the terms of any policy issued in connection with the Contract and shall whenever required produce to the Employer the policy or policies of insurance and the receipts for the payments of the current premiums.

11. Disputes

Settlement of disputes

11.1. If any dispute or difference of any kind whatsoever shall arise between the Employer and the Contractor in connection with or arising out of the Contract or the carrying out of the Works (excluding a dispute under Clause 7.9 but including a dispute as to any act or omission of the Engineer) whether arising during the progress of the Works or after their completion it shall be settled in accordance with the following provisions.

Notice of Dispute

11.2. For the purpose of Clauses 11.3 to 11.5 inclusive a dispute is deemed to arise when one party serves on the other a notice in writing (herein called the Notice of Dispute) stating the nature of the dispute. Provided that no Notice of Dispute may be served unless the party wishing to do so has first taken any step or invoked any procedure available elsewhere in the Contract in connection with the subject matter of such dispute and the other party or the Engineer as the case may be has:-

(a) taken such step as may be required

or

(b) been allowed a reasonable time to take any such action.

Conciliation

11.3. In relation to any dispute notified under Clause 11.2 and in respect of which no Notice to Refer under Clause 11.5 has been served either party may within 28 days of the service of the Notice of Dispute give notice in writing requiring the dispute to be considered under the Institution of Civil Engineers' Conciliation Procedure (1988) or any amendment or modification thereof being in force at the date of such notice and the dispute shall thereafter be referred and considered in accordance with the said Procedure.

Arbitration

11.4. Where a dispute has been referred to a conciliator under the provisions of Clause 11.3 either party may within 28 days of the receipt of the conciliator's recommendation refer the dispute to the arbitration of a person to be agreed upon by the parties by serving on the other party a written Notice to Refer. Where a written Notice to Refer is not served within the said period of 28 days the recommendation of the conciliator shall be deemed to have been accepted in settlement of the dispute.

11.5. Where a dispute has not been referred to a conciliator under the provisions of Clause 11.3 then either party may within 28 days of service of the Notice of Dispute under Clause 11.2 refer the dispute to the arbitration of a person to be agreed upon by the parties by serving on the other party a written Notice to Refer. Where a Notice to Refer is not served within the said period of 28 days the Notice of Dispute shall be deemed to have been withdrawn.

Appointment of arbitrator

11.6. If the parties fail to appoint an arbitrator within 28 days of either party serving on the other party a written Notice to Concur in the appointment of an arbitrator the dispute shall be referred to a person to be appointed on the application of either party by the President (or if he is unable to act by any Vice President) for the time being of the Institution of Civil Engineers.

ICE Arbitration Procedure 1983

11.7. Any such reference to arbitration shall be conducted in accordance with the Institution of Civil Engineers' Arbitration Procedure (1983) or any amendment or modification thereof being in force at the time of the appointment of the arbitrator and unless otherwise agreed in writing shall follow the rules for the Short Procedure in Part F thereof. Such arbitrator shall have full power to open up review and revise any decision instruction direction certificate or valuation of the Engineer.

12. Scotland

Application to Scotland

12.1. If the Works are situated in Scotland the Contract shall in all respects be construed as a Scottish contract and shall be interpreted in accordance with Scots Law and the following Clause shall apply.

12.2. In Clause 11 hereof the word 'arbiter' shall be substituted for the word 'arbitrator' and 'the Institution of Civil Engineers' Arbitration Procedure (Scotland) (1983)' shall be substituted for 'the Institution of Civil Engineers' Arbitration Procedure (1983)'.

Appendix A

The Institution of Civil Engineers Conditions of Contract for Minor Works

Appendix to the Conditions of Contract

(to be prepared before tenders are invited and to be included with the documents supplied to prospective tenderers)

1. Short description of the work to be carried out under the Contract

   ....

   ....

   ....

2. The payment to be made under Article 2 of the Agreement in accordance with Clause 7 will be ascertained on the following basis. (The alternatives not being used are to be deleted. Two or more bases for payment may be used on one Contract.)

   (a) Lump sum
   (b) Measure and value using a priced Bill of Quantities
   (c) Valuation based on a Schedule of Rates (with an indication in the Schedule of the approximate quantities of major items)
   (d) Valuation based on a Daywork Schedule
   (e) Cost plus fee (the cost is to be specifically defined in the Contract and will exclude off-site overheads and profit)

3. Where a Bill of Quantities or a Schedule of Rates is provided the method of measurement used is

   ....

4. Name of the Engineer
   (Clause 2.1) ....

5. Starting date (if known)
   (Clause 4.1) ....

6. Period for completion
   (Clause 4.2) ....

7. Period for completion of parts of the Works (if applicable) and details of the work to be carried out within each such part
   (Clause 4.2)

	Details of work	Period for completion
Part A	....	....
Part B	....	....
Part C	....	....

8. Liquidated damages
   (Clause 4.6) ....

9. Limit of liquidated damages
   (Clause 4.6) ....

10. Defects Correction Period
    (Clause 5.1) ....

11. Rate of retention
    (Clause 7.3) ....

12. Limit of retention
    (Clause 7.3) ....

13. Minimum amount of interim certificate
    (Clause 7.3) ....

14. Bank whose base lending rate is to be used
    (Clause 7.8) ....

15. Insurance of the Works
    (Clause 10.1) Required/Not required

16. Minimum amount of third party insurance (persons and property)
    (Clause 10.6) ....
    Any one accident/Number of accidents unlimited

Appendix A

The Institution of Civil Engineers Conditions of Contract for Minor Works

Notes for guidance

1. This form of contract is intended for use on contracts where:-

(a) the potential risks involved in the Works for both the Employer and the Contractor are adjudged to be small;
(b) the period for completion of the Contract does not exceed 6 months except where the method of payment is on either a daywork or a cost-plus-fee basis;
(c) the Works are of a simple and straightforward nature;
(d) the Contractor has no responsibility for the design of the permanent works other than possibly design of a specialist nature (see Note 6);
(e) the contract value does not exceed £100 000;
(f) the design of the Works, save for any design work for which the Contractor is made responsible (see Note 6), is complete in all essentials before tenders are invited;
(g) nominated sub-contractors are not employed (but see Note 7).

2. The Contract Schedule should list all documents that will form part of the Contract. It is particularly important to ensure that the Appendix to the Conditions of Contract is prepared before tenders are invited (and the Appendix must then be included with the documents supplied to prospective tenderers). Notes on the completion of the Appendix are included in these guidance notes (see Note 13).

3. The method of payment for the Contract should be as stated in the Appendix to the Conditions of Contract but if a Bill of Quantities is used the method of measurement used must also be indicated in the Appendix. If a Daywork Schedule other than the 'Schedules of Daywork carried out incidental to Contract Work' issued by the Federation of Civil Engineering Contractors is to be used the Schedule to be used must be clearly identified in the tender documents.

4. In view of their short duration all contracts should normally be let on a fixed price basis with no provision for price fluctuation. The letting of contracts on a daywork or cost-plus-fee basis is however not precluded (see Note 1(b)).

5. The procedures for letting and administering a minor works contract are intended to be as simple as possible in line with the low risk involved. The Contract should be fully defined in the documents listed in the Contract Schedule. There is no provision for amendment or addition to the Conditions of Contract and this should be avoided.

6. If the Contractor is required to be responsible for design work of a specialist nature which would normally be undertaken by a specialist sub-contractor or supplier (such as structural steelwork, mechanical equipment or an electrical or plumbing installation) full details must be given either in the Specification or in the Appendix to the Conditions of Contract or on the Drawings indicating precisely the Contractor's responsibility in respect of such work.

7. The Engineer may in respect of any work that is to be sub-let or material purchased in connection with the Contract list in the Specification the names of approved sub-contractors or approved suppliers of material. Nothing however should prevent the Contractor carrying out such work himself if he so chooses or from using other sub-contractors or suppliers of his own choice provided their workmanship or product is satisfactory and equal to that from an approved sub-contractor or supplier.

8. It is intended that acceptance should follow within 2 months of the date for submission of tenders.

9. Access as necessary to the Site should be available at the starting date under Clause 4.1.

10. In respect of Clause 2.2 it should be noted that in all normal circumstances the Engineer would not be expected to delegate his powers under Clauses 3.8, 4.4, 5.4 and 7.6.

11. In respect of Clause 3.4 it has to be recognized that in a minor works contract the Contractor might have no full-time supervisor on site and the Contractor may ask for instructions to be delivered or sent elsewhere for the attention of his representative. In these circumstances the Contractor has to accept the fact that urgent instructions might in the interests of safety or for some other reason have to be given directly to the Contractor's operatives on site.

12. The reference to 'any other items' in Clause 7.2 is to permit the Contractor to include in interim valuations other amounts to which he considers himself entitled such as goods or material vested in the Employer but not yet delivered to the Site or the value of temporary works or constructional plant on the Site for which there is separate provision for payment in the Contract.

13. In completing the Appendix to the Conditions of Contract the following points should be noted.

(1) *Clause 2.1 (Name of the Engineer).* It is the intention that the name of the Engineer who will personally be responsible for the Works should be stated.

(2) *Clause 4.2 (Period for completion).* If the Contract requires completion of parts of the Works by specified dates or within specified times such dates or times and details of the work involved in each part must be entered in the Appendix to the Conditions of Contract.

(3) *Clause 4.6 (Liquidated damages and limit of liquidated damages).* A genuine pre-estimate of the likely damage caused by any delay should be assessed and reduced to a daily or weekly rate. The limit of liquidated damages should not exceed 10% of the estimated final contract value and this should be taken into account when assessing the daily or weekly rate.

(4) *Clause 5.1 (Defects Correction Period).* This should normally be 6 months and in no case should exceed 12 months.

(5) *Clause 7.3 (Rate of retention and limit of retention).* The rate of retention should normally be 5%. A limit of retention has to be inserted in the Appendix and this should normally be between the limits of $2\frac{1}{2}$% and 5% of the estimated final Contract value.

(6) *Clause 7.3 (Minimum interim certificate).* It is recommended that the minimum amount of an interim certificate should be 10% of the estimated final Contract value rounded off upwards to the nearest £1000. This minimum only applies up to the date of practical completion of the whole of the Works.

(7) *Clause 10.1 (Insurance of the Works).* This is at the option of the Employer. It must be borne in mind that Contractors frequently carry large excesses on their all-risks policies so that the Contractor then accepts the risk under Clause 10.1 in respect of any uninsured loss. When the insurance under Clause 10.1 is to be provided by the Employer the details of such insurance, including any excesses which the Contractor may be expected to carry, should be stated in the tender documents.

(8) *Clause 10.6 (Third party insurance).* A minimum cover of £500 000 for any one accident/unlimited number of accidents should normally be insisted upon. In certain locations where there is greater risk to adjacent properties a higher limit may be desirable.

14. *Disputes*

(1) The option provided by the Conciliation Procedure under Clause 11.3 is intended to provide a means whereby disputes can be settled with a minimum of delay by obtaining an independent recommendation as to how the matter in dispute should be settled. The conciliation is complete when the conciliator has delivered his recommendations and, if any, his opinion to the parties.

(2) It is normally expected that the party serving a Notice of Dispute under Clause 11.2 will at the same time serve notice in writing either under Clause 11.3 (Conciliation) or Clause 11.5 (Arbitration).

(3) It should be noted that if within the prescribed period of 28 days after service of a Notice of Dispute neither party has made a request in writing for the dispute to be referred to a conciliator nor has served a written Notice to Refer requiring the dispute to be referred to arbitration then the Notice of Dispute becomes void. It is then open to either party to continue the dispute by serving a fresh Notice of Dispute unless the first Notice of Dispute was served within the 28 days allowed under Clause 7.7 (Final certificate) in which case the final certificate becomes final and binding and no further dispute in respect of the Contract is possible.

Appendix B

CLAUSE COMPARISON WITH ICE 5

ICE 5 clauses		Minor works clauses		Comments
1(1)	Definitions			
(a)	Employer	Agreement	Employer	
(b)	Contractor	Agreement	Contractor	
(c)	Engineer	App 4	Engineer	
(d)	Engineer's Representative	2.2	Resident Engineer	
(e)	Contract	1.2	Contract	
(f)	Specification	Schedule	Specification Ref. to be given	
(g)	Drawings	Schedule	Drawing Nrs. to be included	
(h)	Tender Total	—	See Appendices 2 & 3.	Minor works permits lump sum, measure and value, schedule of rates, dayworks or cost-plus-fee as basis of payment.
(i)	Contract Price	—		
(j)	Permanent Works	—		The Works is defined as all work necessary for the completion of the Contract
(k)	Temporary Works	—		
(l)	Works	1.1	Works	
(m)	Section	App 7	Part	
(n)	Site	1.4	Site	Similar definitions
(o)	Construction Plant	—		
(2)	Singular and Plural	—		
(3)	Headings and marginal notes	—		
(4)	Clause References	—		
(5)	Cost	1.3	Cost	In both cases includes overhead costs on and off the Site but not profit
2(1)	Engineer's Representative	2.2 }	Resident Engineer	
(2)	Appointment of assistance	2.2 }		
(3)	Delegation by the Engineer	2.2 }		
(4)	Reference to the Engineer	—		
3	Assignment	8.1	Assignment	
4	Sub-letting	8.2	No sub-letting without Engineer's consent	Which is not to be unreasonably withheld
5	Documents mutually Explanatory	2.3(g)	Engineer's powers to give instructions	
6	Supply of documents	—		No specific number of documents given. Drawings to be complete before tenders are invited [NG 1(f)]
7(1)	Further drawings and	—		
(2)	Notice by Contractor	—		

	ICE 5 clauses		Minor works clauses	Comments
(3)	Delay in issue	—		
(4)	One copy of documents to be kept on site	—		
8(1)	Contractor's General Responsibilities	3.1	Contractor to perform and complete the works	
(2)	Contractor Responsible for Safety of Site Operations	3.5	Setting out and safety of site operations	
9	Contract Agreement	Agreement	Agreement	Optional. Signed under hand
10	Sureties	—		No provision
11(1)	Inspection of Site	—		
(2)	Sufficiency of Tender	—		
12(1)	Adverse Physical Conditions and Obstructions	3.8	Adverse physical conditions	
(2)	Measures to be Taken	2.3(e)	Engineer's powers to give instructions	
(3)	Delay and Extra Cost	6.1 &	Additional payments	
		4.4(e)	Extensions of time	Implied
(4)	Conditions Reasonably Foreseeable	—		
13(1)	Work to satisfaction of Engineer	2.3(d)	Any change in intended sequence	
		(g)	Explanation of any matter	
(2)	Mode and Manner of Construction	—		No requirement to inform Engineer
(3)	Delay and Extra Cost	6.1	Additional payments	
		4.4	Extensions of time	
14(1)	Programme to be Furnished	4.3	Contractor's programme	If requested
(2)	Revised Programme	4.3	Contractor's programme	Revisions may be requested
(3)	Methods of Construction	—		
(4)	Engineer's Consent	—		
(5)	Design Criteria	3.6	Engineer to provide necessary information	
(6)	Delay and Extra Cost	6.1	Engineer to determine additional sums and deductions	
(7)	Responsibility Unaffected by Approval	—		
15(1)	Contractor's Superintendence	3.1	Contractor to perform and complete the Works	
(2)	Contractor's Agent	3.4	Contractor's authorized representative	It is accepted that superintendence will not be full-time [NG 11]

16	Removal of Contractor's Employees	2.3(h)	Engineer's power to give instructions	
17	Setting-out	3.5	Setting out and safety of Site operations	
18	Boreholes and exploratory	2.3(b)	Engineer's power to give instructions	
19(1)	Safety and Security	10.2	Contractor to insure damage to person or property	
(2)	Employer's Responsibilities	10.3		No provision for lighting and watching
20(1)	Care of the Works	10.1	Insurance of the Works	Insurance optional Appendix Item 15
(2)	Responsibility for Reinstatement	—		
(3)	Excepted Risks	1.5	Definitions: Excepted Risks	
21	Insurance of Works, etc.	10.1	Insurance of the Works	
22(1)	Damage to Persons and Property	10.2	Contractor to indemnify against damage to persons, or property	
(2)	Indemnity by the Employer	10.3	Indemnity by Employer	
23(1)	Insurance against Damage to Persons or Property	10.2	Contractor to insure against damage to persons or property	
(2)	Amount and Terms of Insurance	10.6	Employer to approve insurance	
		10.7	Contractor to produce policies of insurance	
24	Accident or Injury to workman	—		
25	Remedy for Contractor's failure to insure	—		
26(1)	Giving of Notices and payment of fees	9.1	Contractor to comply with statutory requirements	
(2)	Contractor to conform with statutes	9.1		
27(1)	Public Utility Street Works Act 1950 Definitions	—		
(2)	Notification by Employer to Contractor	9.3	Contractor's exemption from liability to comply with statutes	
(3)	Service of Notices by Employer	9.2	Employer to obtain consents	
(4)	Notices by Contractor to Employer	—		
(5)	Failure to commence Street Works	—		
(6)	Delays attribution to Variations	6.1	Additional payments	
		4.4(a)	Extension of time for completion	
(7)	Contractor to comply with other obligations of the Act	9.1	Contractor to comply with statutory requirements	

ICE 5 clauses		Minor works clauses		Comments
28(1)	Patent rights	—		
(2)	Royalties	—		
29(1)	Interference with Traffic and Adjoining Property	—		
(2)	Noise and Disturbance	—		Public and common law
30(1)	Avoidance of Damage to Highways	—		
(2)	Transport of Constructional Plant	—		
(3)	Transport of Materials	—		
31(1)	Facilities for other Contractors	3.9	Facilities for other contractors	No payment for facilities
(2)	Delay and Extra cost			
32	Foods, etc.	—		
33	Clearance of site on Completion	—		
34	Rates of wages/hours and Conditions of Operatives	—		
35	Returns of Labour and Plant	—		
36(1)	Quality of Materials and Workmanship and Tests	3.1	Contractor to perform and complete the Works	
(2)	Cost of Samples	2.3(b)	Engineer may instruct tests.	
(3)	Cost of Tests	6.1	Additional payment	
37	Access to Site	3.9	Facilities for other contractors	No provision for access to other work places
38(1)	Examination before covering up	—		No provision for inspection prior to covering up
(2)	Uncovering and making openings	2.3(b)	Engineer's power to give instructions	
39(1)	Removal of Improper Work or Materials	2.3(f)	Engineer's power to give instructions	
(2)	Default of Contractor in compliance	—		
(3)	Failure to disapprove	—		
40(1)	Suspension of the Works	2.6(1)	Engineering may suspend the progress of the Works	Similar
(2)	Suspension lasting more than three months	2.6(2)		Maximum 60 days
41	Commencement of Works	4.1	Starting date to be notified in writing	
42(1)	Possession of Site	—		
(2)	Wayleaves, etc.	4.4(e)	Extension of time for completion	Adequate access to be provided by Employer

43	Time for Completion	4.2	Time for Completion	
44(1)	Extension of time for completion	4.4	Extension of time for completion	Contractor to avoid or minimize delays
(2)	Interim assessment of extension			Decision subject to regular review
(3)	Assessment at Due Date for completion			
(4)	Final Determination of Extension			
45	Night and Sunday Work	—		
46	Rate of Progress	—		
47(1)	Liquided Damages	4.6	Liquidated damages	
(2)	Liquidated Damages for sections	4.6	parts *pro rata*	
(3)	Damage not a Penalty	—		
(4)	Deduction of Liquidated Damages	—		
(5)	Reimbursement of Liquidated Damages	—		
48(1)	Certification of Completion of Works	4.5(1)	Certificate of Completion of Works	
(2)	Completion of sections of occupied parts	(2)	Certificate of Completion of part of Works	
(3)	Completion of other parts of Works			
(4)	Reinstatement of Ground	—		
49(1)	Definition of Period of Maintenance	5.1	Definition of Defects Correction Period	
(2)	Execution of repair work	4.7	Rectification of defects	Contractor to maintain Works until practical completion of whole of Works
(3)	Cost of Execution of Work of repair, etc.	5.2	Cost of remedying defects	
(4)	Remedy on Contractor's Failure to carry out Work Required	5.3	Remedy for Contractor's failure to correct defects	
(5)	Temporary Reinstatement	—		
50	Contractor to search	2.3(b)	Engineer's power to give instructions	
51(1)	Ordered Variations	2.3(a)	Engineer's power to give instructions	
(2)	Ordered Variations to be in writing	2.4	Request for clause under which instruction is issued	
(3)	Changes in Quantity	—		
52(1)	Valuation of Ordered Variations	6.1	Engineer to determine additional sums and deductions	
		6.2	Valuation of additional work	
(2)	Engineer to fix rates	6.1		
(3)	Daywork	2.5	Dayworks	
(4)	Notice of Claims	—		

ICE 5 clauses		Minor works clauses		Comments
53	Vesting goods, material & plant on site	—		Allowance still made for materials on site
54	Vesting goods or materials not on site	—		
55(1)	Quantities	—		
(2)	Correction of errors	2.3(g)	Engineer to explain	Bills warranted to be in accordance with method of measurement [App. 3]
56(1)	Measurement and Valuation	7.1	Valuation of the Works	Payment in accordance with App. 2.
(2)	Increase or decrease of rates			
(3)	attending for measurement	6.1	Engineer to determine rate	
57	Method of Measurement	App.3	Method of measurement used	
58(1)	Provisional sums	—		No provision for provisional sums, prime cost items, or nominated sub-contractors. Express requirement for specialist design [NG6]
(2)	Prime cost items	—		
(3)	Design Requirements to be expressly required			
(4)	Use of Prime Cost items			
(5)	Nominated Sub-contractors: definition	—		
(6)	Production of Vouchers	—		
(7)	Use of Provisional Sums	—		
59A				
59B	Nominated Sub-contractors	—		
59C				
60(1)	Monthly statement	7.3	Monthly statements	
(2)	Monthly payments	7.3	Interim payments	
(3)	Final account	7.6	Contractor to submit final account	
		7.7	Final Certificate conclusive amount due	No provision in ICE 5
		7.7	Disputing final account	Normal C1 66 procedure in ICE 5
(4)	Retention	7.3	Retention	
(5)	Payment of retention	App's 11 & 12	Payment of retention	
		7.4	Payment of 1st half	
		7.5	Payment of 2nd half	
(6)	Interest on overdue accounts	7.8	Interest on overdue payments	
(7)	Correction & Withholding Certificates	7.3	Interim payments	
(8)	Copy Certificate for Contractor	—		

61(1)	Maintenance Certificate	5.4	Engineer to certify completion	
(2)	Unfulfilled obligations	5.5	Unfulfilled obligations	
62	Urgent repairs	—		
63(1)	Forfeiture	—		No provision
(2)	Assignment to Employer	—		
(3)	Valuation at Date of Forfeiture	—		
64	Payment in Event of Frustration	—		Payment possible Common law
65(1)	Works to continue for 28 days on outbreak of war	—		
(2)	Effect of Completion within 28 days	—		
(3)	Right of Employer to determine Contract	—		
(4)	Removal of Plant on Determination	—		
(5)	Payment on Determination	—		
(6)	Provisions to apply as from outbreak of War	—		
66(1)	Settlement of Disputes Arbitration	11.1	Settlement of Disputes	
(2)	Engineer's Decision — Effect on Contractor and Employer	11.3	Conciliation	Engineer's decision replaced by conciliation
(3)	Time for Engineer's decision	11.4 & 5	Arbitration	
(4)	President or Vice President to act	11.6	Appointment of arbitrator	
(5)	ICE Arbitration Procedure	11.7	ICE Arbitration Procedure (1983)	Short Procedure F unless otherwise agreed
(6)	Engineer as Witness	—		
76(1)	Application to Scotland	12.1	Application to Scotland	
(2)	Ditto Arbitration	12.2		
68(1)	Service of Notices on Contractor	—		
(2)	Service of Notices on Employer	—		
69	Tax Fluctuations	—		
70	Value Added Tax	7.9	Value-added tax	
71	Metrication	—		
72	Additional Clauses	—	No provision for amendment	

Appendix C

ICE DISPUTE RESOLUTION

ICE Conciliation Procedure 1988 and
ICE Arbitration Procedure (1983)

The Institution of Civil Engineers Conciliation Procedure 1988

Conciliation shall be carried out in accordance with the following rules.

1. These rules are deemed to have been agreed between the parties to apply whenever they have entered into a contract which provides for conciliation for any dispute or difference which may arise between parties in accordance with the Institution of Civil Engineers' Conciliation Procedure 1988 or where the parties agreed that conciliation will apply in accordance with that procedure.

2. These rules shall be interpreted and applied in the manner most conducive to the efficient conduct of the proceedings with the primary objective of obtaining the conciliator's recommendation as quickly as possible.

3. Subject to the provisions of the Contract relating to conciliation until such time as the Contractor has completed all his obligations under the Contract either party to the Contract may by Notice in writing to the other party or parties request that any dispute, difference or other matter in connection with or arising out of the Contract or the carrying out of the Works shall be referred to a conciliator for his recommendation. Such Notice shall be accompanied by a brief statement of the matter or matters upon which it is desired to receive the conciliator's recommendation and the relief and remedy sought.

4. Save where a conciliator has already been appointed, the parties shall agree a conciliator within 14 days of the Notice being given under Rule 3. In default of agreement any party may request the President (or, if he is unable to act, any Vice President) for the time being of the Institution of Civil Engineers to appoint a conciliator within 14 days of receipt of the request by him which request shall be accompanied by a copy of the Notice.

5. The party requesting conciliation shall deliver to the conciliator upon his appointment under Rule 3 or 4 a copy of the Notice prescribed by Rule 3 together with the names and addresses of the parties' representatives.

6. The conciliator shall start the conciliation as soon as possible after his appointment and shall use his best endeavours to conclude the conciliation as soon as possible, and in any event within 2 months of his appointment unless the parties otherwise agree.

7. Any party may upon receipt of the Notice under Rule 3 or the appointment of the conciliator under Rule 4 (whichever shall be the later) and within such period as the conciliator may allow send to the conciliator and each other written submissions stating their version of the dispute, difference or other matter together with their views as to the rights and liabilities of the parties arising from it and the financial consequences. Copies of all relevant documents relied on shall be attached to any written submission which may be accompanied by written statement of evidence.

8. With the prior agreement of the conciliator a further period not exceeding 14 days shall be allowed after the period allowed by the conciliator for written submissions under Rule 7 during which any party may send a further written submission to the conciliator and each other replying specifically to points made in any other party's original submission.

9. The conciliator may on his own initiative at any time after his appointment and upon giving not less than 24 hours' notice to the parties visit and inspect the Site or the subject matter of the dispute. He may generally inform himself in any way he thinks fit of the nature and

facts of the dispute, difference or other matter referred to him, including meeting the parties separately.

10. The conciliator may convene a meeting at which the parties shall be present. He shall give the parties not less than 7 days' notice of such a meeting unless they agree a shorter period. At the meeting the conciliator may take evidence and hear submissions on behalf of any party but shall not be bound by the rules of evidence or by any rules of procedure other than these rules. If it is not possible to conclude the business of any meeting held under this rule on the day or days appointed by him the conciliator may adjourn such meeting to a day to be fixed by him.

11. The conciliator may and shall if requested by all parties seek legal advice or other advice.

12. (1) Subject to Rules 6, 7 and 8 and within 21 days of the conclusion of any meeting held pursuant to Rule 10 the conciliator shall prepare his recommendation as to the way in which the matter shall be disposed of and settled between the parties including any recommendation as to any sum of money which should be paid by one party to any other party.

(2) If the conciliator considers it appropriate so to do he may at the same time or within 7 days of the giving of his recommendation also submit in a separate document his written opinion on the matter or on any part of the matter referred to him. The conciliator's opinion, if given, shall contain such reasons for and comments thereon as in all the circumstances he may deem appropriate.

(3) The conciliator may at any time at his discretion, if he considers it appropriate, or is so requested by the parties, express his preliminary views on the matter referred to him.

13. When the conciliator has prepared his recommendation he shall notify the parties in writing and send them an account of his fees and disbursements. Unless otherwise agreed between themselves each party shall be responsible for paying and shall within 7 days of receipt of notice from the conciliator pay an equal share of the account save that the parties shall be jointly and severally liable to the conciliator for the whole of his account. Upon receipt of payment in full the conciliator shall send his recommendation to all the parties provided that if any party shall fail to make the payment due from him any other party may after giving 7 days' notice in writing to the defaulting party pay the sum to the conciliator and recover the amount from the defaulting party as a debt due.

14. The conciliator shall not be appointed arbitrator in any subsequent arbitration between the parties whether arising out of the dispute, difference or other matter or otherwise arising out of the same Contract unless the parties otherwise agree in writing.

15. Any document required by these rules shall be sent to the parties by recorded delivery to the principal place of business or if a company to its registered office. Any document required by these rules to be sent to the conciliator shall be sent by recorded delivery to him at the address which he shall notify to the parties on his appointment.

The Institution of Civil Engineers'
ARBITRATION PROCEDURE (1983)

Part A. Reference and Appointment

Rule 1. Notice to Refer

1.1 A dispute or difference shall be deemed to arise when a claim or assertion made by one party is rejected by the other party and that rejection is not accepted. Subject only to Clause 66(1) of *the ICE Conditions of Contract* (if applicable) either party may then invoke arbitration by serving a *Notice to Refer* on the other party.

1.2 The Notice to Refer shall list the matters which the issuing party wishes to be referred to arbitration. Where Clause 66 of the ICE Conditions of Contract applies the Notice to Refer shall also state the date when the matters listed therein were referred to the Engineer for his decision under Clause 66(1) and the date on which the Engineer gave his decision thereon or that he has failed to do so.

Rule 2. Appointment of sole Arbitrator by agreement

2.1 After serving the Notice to Refer either party may serve upon the other a *Notice to Concur* in the appointment of an Arbitrator listing therein the names and addresses of any persons he proposes as Arbitrator.

2.2 Within 14 days thereafter the other party shall

(*a*) agree in writing to the appointment of one of the persons listed in the Notice to Concur

or (*b*) propose a list of alternative persons

2.3 Once agreement has been reached the issuing party shall write to the person so selected inviting him to accept the appointment enclosing a copy of the Notice to Refer and documentary evidence of the other party's agreement.

2.4 If the person so selected accepts the appointment he shall notify the issuing party in writing and send a copy to the other party. The date of posting or service as the case may be of this notification shall be deemed to be the date on which the Arbitrator's appointment is completed.

Rule 3. Appointment of sole Arbitrator by the President

3.1 If within one calendar month from service of the Notice to Concur the parties fail to appoint an Arbitrator in accordance with Rule 2 either party may then apply to the President to appoint an Arbitrator. The parties may also agree to apply to the President without a Notice to Concur.

3.2 Such application shall be in writing and shall include copies of the Notice to Refer, the Notice to Concur (if any) and any other relevant documents. The application shall be accompanied by the appropriate fee.

3.3 The Institution will send a copy of the application to the other party stating that the President intends to make the appointment on a specified date. Having first contacted an appropriate person and obtained his agreement the President will make the appointment on the specified date or such later date as may be appropriate which shall then be deemed to be the date on which the Arbitrator's appointment is completed. The Institution will notify both parties and the Arbitrator in writing as soon as possible thereafter.

Rule 4. Notice of further disputes or differences

4.1 At any time before the Arbitrator's appointment is completed either party may put forward further disputes or differences to be referred to him. This shall be done by serving upon the other party an additional Notice to Refer in accordance with Rule 1.

4.2 Once his appointment is completed the Arbitrator shall have jurisdiction over any issue connected with and necessary to the determination of any dispute or difference already referred to him whether or not the connected issue has first been referred to the Engineer for his decision under Clause 66(1) of the ICE Conditions of Contract.

Part B. Powers of the Arbitrator

Rule 5. Power to control the proceedings

5.1 The Arbitrator may exercise any or all of the powers set out or necessarily to be implied in this Procedure on such terms as he thinks fit. These terms may include orders as to costs, time for compliance and the consequences of non-compliance.

5.2 Powers under this Procedure shall be in addition to any other powers available to the Arbitrator.

Rule 6. Power to order protective measures

6.1 The arbitrator shall have power

(*a*) to give directions for the detention storage sale or disposal of the whole or any part of the subject matter of the dispute at the expense of one or both of the parties

(*b*) to give directions for the preservation of any document or thing which is or may become evidence in the arbitration

(*c*) to order the deposit of money or other security to secure the whole or any part of the amount(s) in dispute

(*d*) to make an order for security for costs in favour of one or more of the parties

and (*e*) to order his own costs to be secured

6.2 Money ordered to be paid under this Rule shall be paid without delay into a separate bank account in the name of a stakeholder to be appointed by and subject to the directions of the Arbitrator.

Rule 7. Power to order concurrent Hearings

7.1 Where disputes or differences have arisen under two or more contracts each concerned wholly or mainly with the same subject matter and the resulting arbitrations have been referred to the same Arbitrator he may with the agreement of all the parties concerned or upon the application of one of the parties being a party to all the contracts involved order that the whole or any part of the matters at issue shall be heard together upon such terms or conditions as the Arbitrator thinks fit.

7.2 Where an order for concurrent Hearings has been made under Rule 7.1 the Arbitrator shall nevertheless make and publish separate Awards unless the parties otherwise agree but the Arbitrator may if he thinks fit prepare one combined set of Reasons to cover all the Awards.

Rule 8. Powers at the Hearing

8.1 The Arbitrator may hear the parties their representatives and/or witnesses at any time or place and may adjourn the arbitration for any period on the application of any party or as he thinks fit.

8.2 Any party may be represented by any person including in the case of a company or other legal entity a director officer employee or beneficiary of such company or entity. In particular, a person shall not be prevented from representing a party because he is or may be also a witness in the proceedings. Nothing shall prevent a party from being represented by different persons at different times.

8.3 Nothing in these Rules or in any other rule custom or practice shall prevent the Arbitrator from starting to hear the arbitration once his appointment is completed or at any time thereafter.

8.4 Any meeting with or summons before the Arbitrator at which both parties are represented shall if the Arbitrator so directs be treated as part of the hearing of the arbitration.

Rule 9. Power to appoint assessors or to seek outside advice

9.1 The Arbitrator may appoint a legal technical or other assessor to assist him in the conduct of the arbitration. The Arbitrator shall direct when such assessor is to attend hearings of the arbitration.

9.2 The Arbitrator may seek legal technical or other advice on any matter arising out of or in connection with the proceedings.

9.3 Further and/or alternatively the Arbitrator may rely upon his own knowledge and expertise to such extent as he thinks fit.

Part C. Procedure before the Hearing

Rule 10. The preliminary meeting

10.1 As soon as possible after accepting the appointment the Arbitrator shall summon the parties to a preliminary meeting for the purpose of giving such directions about the procedure to be adopted in the arbitration as he considers necessary.

10.2 At the preliminary meeting the parties and the Arbitrator shall consider whether and to what extent

(*a*) Part F (Short Procedure) or Part G (Special Procedure for Experts) of these Rules shall apply

(*b*) the arbitration may proceed on documents only

(*c*) progress may be facilitated and costs saved by determining some of the issues in advance of the main Hearing

(*d*) the parties should enter into an exclusion agreement (if they have not already done so) in accordance with S.3 of the Arbitration Act 1979 (where the Act applies to the arbitration)

and in general shall consider such other steps as may minimise delay and expedite the determination of the real issues between the parties.

10.3 If the parties so wish they may themselves agree directions and submit them to the Arbitrator for his approval. In so doing the parties shall state whether or not they wish Part F or Part G of these Rules to apply. The Arbitrator may then approve the directions as submitted or (having first consulted the parties) may vary them or substitute his own as he thinks fit.

Rule 11. Pleadings and discovery

11.1 The Arbitrator may order the parties to deliver pleadings or statements of their cases in any form he thinks appropriate. The Arbitrator may order any party to answer the other party's case and to give reasons for any disagreement.

11.2 The Arbitrator may order any party to deliver in advance of formal discovery copies of any documents in his possession custody or power which relate either generally or specifically to matters raised in any pleading statement or answer.

11.3 Any pleading statement or answer shall contain sufficient detail for the other party to know the case he has to answer. If sufficient detail is not provided the Arbitrator may of his own motion or at the request of the other party order further and better particulars to be delivered.

11.4 If a party fails to comply with any order made under this Rule the Arbitrator shall have power to debar that party from relying on the matters in respect of which he is in default and the Arbitrator may proceed with the arbitration and make his Award accordingly. Provided that the Arbitrator shall first give notice to the party in default that he intends to proceed under this Rule.

Rule 12. Procedural meetings

12.1 The Arbitrator may at any time call such procedural meetings as he deems necessary to identify or clarify the issues to be decided and the procedures to be adopted. For this purpose the Arbitrator may request particular persons to attend on behalf of the parties.

12.2 Either party may at any time apply to the Arbitrator for leave to appear before him on any interlocutory matter. The Arbitrator may call a procedural meeting for the purpose or deal with the application in correspondence or otherwise as he thinks fit.

12.3 At any procedural meeting or otherwise the Arbitrator may give such directions as he thinks fit for the proper conduct of the arbitration. Whether or not formal pleadings have been ordered under Rule 11 such directions may include an order that either or both parties shall prepare in writing and shall serve upon the other party and the Arbitrator any or all of the following

(*a*) a summary of that party's case

(*b*) a summary of that party's evidence

(*c*) a statement or summary of the issues between the parties

(*d*) a list and/or a summary of the documents relied upon

(*e*) a statement or summary of any other matters likely to assist the resolution of the disputes or differences between the parties

Rule 13. Preparation for the Hearing

13.1 In addition to his powers under Rules 11 and 12 the Arbitrator shall also have power

(*a*) to order that the parties shall agree facts as facts and figures as figures where possible

(*b*) to order the parties to prepare an agreed bundle of all documents relevant to the arbitration. The agreed bundle shall thereby be deemed to have been entered in evidence without further proof and without being read out at the Hearing. Provided always that either party may at the Hearing challenge the admissibility of any document in the agreed bundle.

(*c*) to order that any experts whose reports have been exchanged before the Hearing shall be examined by the Arbitrator in the presence of the parties or their legal representatives and not by the parties or their legal representatives themselves. Where such an order is made either party may put questions whether by way of cross-examination or re-examination to any party's expert after all experts have been examined by the Arbitrator provided that the party so doing shall first give notice of the nature of the questions he wishes to put.

13.2 Before the Hearing the Arbitrator may and shall if so requested by the parties read the documents to be used at the Hearing. For this or any other purpose the Arbitrator may require all such documents to be delivered to him at such time and place as he may specify.

Rule 14. Summary Awards

14.1 The Arbitrator may at any time make a *Summary Award* and for this purpose shall have power to award payment by one party to another of a sum representing a reasonable proportion of the final nett amount which in his opinion that party is likely to be ordered to pay after determination of all the issues in the arbitration and after taking into account any defence or counterclaim upon which the other party may be entitled to rely.

14.2 The Arbitrator shall have power to order the party against whom a Summary Award is made to pay part or all of the sum awarded to a stakeholder. In default of compliance with such an order the Arbitrator may order payment of the whole sum in the Summary Award to the other party.

14.3 The Arbitrator shall have power to order payment of costs in relation to a Summary Award including power to order that such costs shall be paid forthwith.

14.4 A Summary Award shall be final and binding upon the parties unless and until it is varied by any subsequent Award made and published by the same Arbitrator or by any other arbitrator having jurisdiction over the matters in dispute. Any such subsequent Award may order repayment of monies paid in accordance with the Summary Award.

Part D. Procedure at the Hearing

Rule 15. The Hearing

15.1 At or before the Hearing and after hearing representations on behalf of each party the Arbitrator shall determine the order in which the parties shall present their cases and/or the order in which the issues shall be heard and determined.

15.2 The Arbitrator may order any submission or speech by or on behalf of any party to be put into writing and delivered to him and to the other party. A party so ordered shall be entitled if he so wishes to enlarge upon or vary any such submission orally.

15.3 The Arbitrator may on the application of either party or of his own motion hear and determine any issue or issues separately.

15.4 If a party fails to appear at the Hearing and provided that the absent party has had notice of the Hearing or the Arbitrator is satisfied that all reasonable steps have been taken to notify him of the Hearing the Arbitrator may proceed with the Hearing in his absence. The Arbitrator shall nevertheless take all reasonable steps to ensure that the real issues between the parties are determined justly and fairly.

Rule 16. Evidence

16.1 The Arbitrator may order a party to submit in advance of the Hearing a list of the witnesses he intends to call. That party shall not thereby be bound to call any witness so listed and may add to the list so submitted at any time.

16.2 No expert evidence shall be admissible except by leave of the Arbitrator. Leave may be given on such terms and conditions as the Arbitrator thinks fit. Unless the Arbitrator otherwise orders such terms shall be deemed to include a requirement that a report from each expert containing the substance of the evidence to be given shall be served upon the other party within a reasonable time before the Hearing.

16.3 The arbitrator may order disclosure or exchange of proofs of evidence relating to factual issues. The Arbitrator may also order any party to prepare and disclose in advance a list of points or questions to be put in cross-examination of any witness.

16.4 Where a list of questions is disclosed whether pursuant to an order of the Arbitrator or otherwise the party making disclosure shall not be bound to put any question therein to the witness unless the Arbitrator so orders. Where the party making disclosure puts a question not so listed in cross-examination the Arbitrator may disallow the costs thereby occasioned.

16.5 The Arbitrator may order that any proof of evidence which has been disclosed shall stand as the evidence in chief of the deponent provided that the other party has been or will be given an opportunity to cross-examine the deponent thereon. The Arbitrator may also at any time before such cross-examination order the deponent or some other identified person to deliver written answers to questions arising out of the proof of evidence.

16.6 The Arbitrator may himself put questions to any witness and/or require the parties to conduct enquiries tests or investigations. Subject to his agreement the parties may ask the Arbitrator to conduct or arrange for any enquiry test or investigation.

Part E. After the Hearing

Rule 17. The Award

17.1 Upon the closing of the Hearing (if any) and after having considered all the evidence and submissions the Arbitrator will prepare and publish his Award.

17.2 When the Arbitrator has made and published his Award (including a Summary Award under Rule 14) he will so inform the parties in writing and shall specify how and where it may be taken up upon due payment of his fee.

Rule 18. Reasons

18.1 Whether requested by any party to do so or not the Arbitrator may at his discretion state his Reasons for all or any part of his Award. Such Reasons may form part of the Award itself or may be contained in a separate document.

18.2 A party asking for Reasons shall state the purpose for his request. If the purpose is to use them for an appeal (whether under S.1 of the Arbitration Act 1979 or otherwise) the requesting party shall also specify the points of law with which he wishes the Reasons to deal. In that event the Arbitrator shall give the other party an opportunity to specify additional points of law to be dealt with.

18.3 Reasons prepared as a separate document may be delivered with the Award or later as the Arbitrator thinks fit.

18.4 Where the Arbitrator decides not to state his Reasons he shall nevertheless keep such notes as will enable him to prepare Reasons later if so ordered by the High Court.

Rule 19. Appeals

19.1 If any party applies to the High Court for leave to appeal against any Award or decision or for an order staying the arbitration proceedings or for any other purpose that party shall forthwith notify the Arbitrator of the application.

19.2 Once any Award or decision has been made and published the Arbitrator shall be under no obligation to make any statement in connection therewith other than in compliance with an order of the High Court under S.1(5) of the Arbitration Act 1979.

Part F. Short Procedure

Rule 20. Short Procedure

20.1 Where the parties so agree (either of their own motion or at the invitation of the Arbitrator) the arbitration shall be conducted in accordance with the following *Short Procedure*.

20.2 Each party shall set out his case in the form of a file containing

(*a*) a statement as to the orders or awards he seeks

(*b*) a statement of his reasons for being entitled to such orders or awards

and (*c*) copies of any documents on which he relies (including statements) identifying the origin and date of each document

and shall deliver copies of the said file to the other party and to the Arbitrator in such manner and within such time as the Arbitrator may direct.

20.3 After reading the parties' cases the Arbitrator may view the site or the Works and may require either or both parties to submit further documents or information in writing.

20.4 Within one calendar month of completing the foregoing steps the Arbitrator shall fix a day when he shall meet the parties for the purpose of

(*a*) receiving any oral submissions which either party may wish to make

and/or (*b*) the Arbitrator's putting questions to the parties their representatives or witnesses

For this purpose the Arbitrator shall give notice of any particular person he wishes to question but no person shall be bound to appear before him.

20.5 Within one calendar month following the conclusion of the meeting under Rule 20.4 or such further period as the Arbitrator may reasonably require the Arbitrator shall make and publish his Award.

Rule 21. Other matters

21.1 Unless the parties otherwise agree the Arbitrator shall have no power to award costs to either party and the Arbitrator's own fees and charges shall be paid in equal shares by the parties. Where one party has agreed to the Arbitrator's fees the other party by agreeing to this Short Procedure shall be deemed to have agreed likewise to the Arbitrator's fees.

21.2 Either party may at any time before the Arbitrator has made and published his Award under this Short Procedure require by written notice served on the Arbitrator and the other party that the arbitration shall cease to be conducted in accordance with this Short Procedure. Save only for Rule 21.3 the Short Procedure shall thereupon no longer apply or bind the parties but any evidence already laid before the Arbitrator shall be admissible in further proceedings as if it had been submitted as part of those proceedings and without further proof.

21.3 The party giving written notice under Rule 21.2 shall thereupon in any event become liable to pay

(*a*) the whole of the Arbitrator's fees and charges incurred up to the date of such notice

and (*b*) a sum to be assessed by the Arbitrator as reasonable compensation for the costs (including any legal costs) incurred by the other party up to the date of such notice.

Payment in full of such charges shall be a condition precedent to that party's proceeding further in the arbitration unless the Arbitrator otherwise directs. Provided that non-payment of the said charges shall not prevent the other party from proceeding in the arbitration.

Part G. Special Procedure for Experts

Rule 22. Special Procedure for Experts

22.1 Where the parties so agree (either of their own motion or at the invitation of the Arbitrator) the hearing and determination of any issues of fact which depend upon the evidence of experts shall be conducted in accordance with the following *Special Procedure.*

22.2 Each party shall set out his case on such issues in the form of a file containing

(*a*) a statement of the factual findings he seeks

(*b*) a report or statement from and signed by each expert upon whom that party relies

and (*c*) copies of any other documents referred to in each expert's report or statement or on which the party relies identifying the origin and date of each document

and shall deliver copies of the said file to the other party and to the Arbitrator in such manner and within such time as the Arbitrator may direct.

22.3 After reading the parties' cases the Arbitrator may view the site or the Works and may require either or both parties to submjt further documents or information in writing.

22.4 Thereafter the Arbitrator shall fix a day when he shall meet the experts whose reports or statements have been submitted. At the meeting each expert may address the Arbitrator and put questions to any other expert representing the other party. The Arbitrator shall so direct the meeting as to ensure that each expert has an adequate opportunity to explain his opinion and to comment upon any opposing opinion. No other person shall be entitled to address the Arbitrator or question any expert unless the parties and the Arbitrator so agree.

22.5 Thereafter the Arbitrator may make and publish an Award setting out with such details or particulars as may be necessary his decision upon the issues dealt with.

Rule 23. Costs

23.1 The Arbitrator may in his Award make orders as to the payment of any costs relating to the foregoing matters including his own fees and charges in connection therewith.

23.2 Unless the parties otherwise agree and so notify the Arbitrator neither party shall be entitled to any costs in respect of legal representation assistance or other legal work relating to the hearing and determination of factual issues by this Special Procedure.

Part H. Interim Arbitration

Rule 24. Interim Arbitration

24.1 Where the Arbitrator is appointed and the arbitration is to proceed before completion or alleged completion of the Works then save in the case of a dispute arising under Clause 63 of the ICE Conditions of Contract the following provisions shall apply in addition to the foregoing Rules and the arbitration shall be called an Interim Arbitration.

24.2 In conducting an Interim Arbitration the Arbitrator shall apply the powers at his disposal with a view to making his Award or Awards as quickly as possible and thereby allowing or facilitating the timely completion of the Works.

24.3 Should an Interim Arbitration not be completed before the Works or the relevant parts thereof are complete the Arbitrator shall within 14 days of the date of such completion make and publish his Award findings of fact or Interim Decision pursuant to Rule 24.5 hereunder on the basis of evidence given and submissions made up to that date together with such further evidence and submissions as he may in his discretion agree to receive during the said 14 days. Provided that before the expiry of the said 14 days the parties may otherwise agree and so notify the Arbitrator.

24.4 For the purpose only of Rule 24.3 the Arbitrator shall decide finally whether and if so when the Works or the relevant parts thereof are complete.

24.5 In an Interim Arbitration the Arbitrator may make and publish any or all of the following

(*a*) a Final Award or an Interim Award on the matters at issue therein

(*b*) findings of fact

(*c*) a Summary Award in accordance with Rule 14

(*d*) an Interim Decision as defined in Rule 24.6.

An Award under (*a*) above or a Finding under (*b*) above shall be final and binding upon the parties in any subsequent proceedings. Anything not expressly identified as falling under either of headings (*a*) (*b*) or (*c*) above shall be deemed to be an Interim Decision under heading (*d*). Save as aforesaid the Arbitrator shall not make an Interim Decision without first notifying the parties that he intends to do so.

24.6 An *Interim Decision* shall be final and binding upon the parties and upon the Engineer (if any) until such time as the Works have been completed or any Award or decision under Rule 24.3 has been given. Thereafter the Interim Decision may be re-opened by another Arbitrator appointed under these Rules and where such other Arbitrator was also the Arbitrator appointed to conduct the Interim Arbitration he shall not be bound by his earlier Interim Decision.

24.7 The Arbitrator in an Interim Arbitration shall have power to direct that Part F (Short Procedure) and/or Part G (Special Procedure for Experts) shall apply to the Interim Arbitration.

Part J. Miscellaneous

Rule 25. Definitions

25.1 In these Rules the following definitions shall apply.

(*a*) 'Arbitrator' includes a tribunal of two or more Arbitrators or an Umpire.

(*b*) 'Institution' means The Institution of Civil Engineers.

(*c*) 'ICE Conditions of Contract' means the Conditions of Contract for use in connection with Works of Civil Engineering Construction published jointly by the Institution, the Association of Consulting Engineers and the Federation of Civil Engineering Contractors.

(*d*) 'Other party' includes the plural unless the context otherwise requires.

(*e*) 'President' means the President for the time being of the Institution or any Vice-President acting on his behalf.

(*f*) 'Procedure' means The Institution of Civil Engineers' Arbitration Procedure (1983) unless the context otherwise requires.

(*g*) 'Award', 'Final Award' and 'Interim Award' have the meanings given to those terms in or in connection with the Arbitration Acts 1950 to 1979. 'Summary Award' means an Award made under Rule 14 hereof.

(*h*) 'Interim Arbitration' means an arbitration in accordance with Part H of these Rules. 'Interim Decision' means a decision as defined in Rule 24.6 hereof.

Rule 26. Application of the ICE Procedure

26.1 This Procedure shall apply to the conduct of the arbitration if

(*a*) the parties at any time so agree

(*b*) the President when making an appointment so directs

or (*c*) the Arbitrator so stipulates at the time of his appointment

Provided that where this Procedure applies by virtue of the Arbitrator's stipulation under (*c*) above the parties may within 14 days of that appointment agree otherwise in which event the Arbitrator's appointment shall terminate and the parties shall pay his reasonable charges in equal shares.

26.2 This Procedure shall not apply to arbitrations under the law of Scotland for which a separate *ICE Arbitration Procedure (Scotland)* is available.

26.3 Where an arbitration is governed by the law of a country other than England and Wales this Procedure shall apply to the extent that the applicable law permits.

Rule 27. Exclusion of liability

27.1 Neither the Institution nor its servants or agents nor the President shall be liable to any party for any act omission or misconduct in connection with any appointment made or any arbitration conducted under this Procedure.

STATUTES AND CASES CITED IN TEXT

Statutes

Statute	Page
Arbitration Acts 1934, 1950, 1979	137,146,148,184
Companies Act 1989	18
Control of Pollution Act 1974	121
Courts and Legal Services Act 1990	183
Employer's Liability (Compulsory Insurance) Act 1969	125
Health and Safety at Work, etc. Act 1974	53,169
Highways Act 1980	121
Latent Damage Act 1986	84
Law of Property (Miscellaneous Provisions) Act 1989	18
Limitations Act 1980	18,82
New Roads and Street Works Act 1991	121–123
Public Health Act 1961	121
Public Utility Street Works Act 1950	96,121–123
Supply of Goods and Services Act 1982	46,60,160
Value Added Tax Act 1983	163

Cases

Case	Page
Appleby v *Myers (1867)* LR 2 CP 651	184
D & F Estates v *Church Commissioners* (1988) 3 WLR 368	46
Davis Contractors v *Fareham UDC* (1956) AC 696	27
Derek Crouch Construction Co. Ltd v *Nothern Regional Health Authority* (1983) 24 BLR 60	184
Glenlion Construction Ltd v *The Guinness Trust* (1987) 30 BLR 89	46, 180
Greater Nottingham Co-operative Society Ltd v *Cementation Foundations and Engineering Ltd* (1988) 33 BLR 43 CA	46
Greaves v *Baynham Meikle* (1972) 1 WRL 1095	46
Hair v *Prudential Insurance Co. Ltd* (1982) 6 ILR 104	135
Hedley Byrne v *Heller & Partners* (1964) AC 465	47
Hickman & Co v *Roberts* (1913) AC 299	48
Holland Dredging (UK) Ltd v *The Dredging and Construction Co. Ltd and ICI* (1987) 37 BLR 1	184
Holland Hannen & Cubbits(Nothern) Ltd v *Welsh Health Technical Services Organisation* (1985) 35 BLR 1 CA	48
IBA v *EMI* 14 BLR 1	46
Lindenau v *Desborough* (1828) 8 B & C 586	134
Lister v *Romford Ice and Cold Storage Company* (1957) AC 555	46
Morgan-Grenfell (Local Authority Finance) Ltd v *Seven Seas Dredging Ltd* (No. 2) (1990)	107–108
Murphy v *Brentwood District Council* (HL 1990)	46
Neodox v *The Borough of Swinton and Pendlebury* (1958), 5 BLR 34	47
Pacific Associates v *Baxter* (1989) 44 BLR 33	31, 47, 48
Pearce (CJ) v *Hereford Corportation* (1967) 66 LGR 647	184
Perini Corporation v *Commonwealth of Australia* (1969) 12 BLR 82	48
Sudbrooke Trading Estate v *Eggleston* (1982) 1 All ER 1, (1983) 1 AC 444	88
Sutcliffe v *Thackrah* (1974) 1 All ER 859	47
Thomas Feather (Bradford) Ltd v *Keighley Corporation* (1953) 52 LGR 30	112
Trollope & Colls Ltd v *Atomic Power Constructors Ltd* (1962) 3 All ER 1035	17
Viking Grain Storage Ltd v *T.H.White Installations* (1985) 33 BLR 103	46
Yorkshire Water Authority v *Sir Alfred McAlpine & Son (Northern) Ltd* (1985) 32 BLR 114	28, 38
Young and Martin v *McManus Childs* (1969) 1AC 454	46

Appendix E

SPECIMEN FORMS

APPENDIX E

Contents

1. Courtesy of the Federation of Civil Engineering Contractors.

1. TENDERS

Letter of invitation to tender

Consulting Engineers

37-39 High Street
East Cheam
Avon
BS27 8QD

McTar Ltd
Surfacing Contractors
23 High Road
Byway
Essex

23 April 1992

Dear Sirs

Re: Gallows Estate Roads Contract 1

On behalf of our client, E. M. Ployer Limited, we invite you to tender for these works in accordance with the accompanying documents.

Should you be unable or not wish to tender please return the documents to us as soon as possible so that we may invite another contractor to tender.

We would point out that our client does not undertake to accept the lowest or any tender.

Yours faithfully

T. Telford

Instructions to tenderers

1. The Contract does not include a price fluctuation clause, and the rates and prices will be fixed for the duration of the Contract.

2. You should return with your tender the following documents:

(i) the form of tender,
(ii) the Contract Schedule,
(iii) the Appendix to the Conditions of Contract,
(iv) the priced Bill of Quantities/Schedule of Rates.

3. The insurance requirements are detailed in Clause 2 of the Special Conditions of Contract. The Contractor will be required to provide proof to the Engineer that the required insurances have been obtained before starting work upon the Site.

4. Any queries that arise during the tender period will be dealt with by Mr D. Brunel (tel: 010 693241).

5. Tenders must be submitted strictly in accordance with these enclosed documents. Any amendments or qualifications made by the Contractor may result in the tender being rejected.

6. No alternative proposals will be considered.

7. If any errors occur between the rates and the extensions, the rates will be accepted as being correct. The Engineer will make any necessary correction and will require the Contractor either to confirm all such corrections or withdraw his tender.

9. The Engineer will inform all tenderers within 7 days of the opening of tenders the names of the tenderers who submitted tenders in alphabetical order, together with a list of the tendered sums in descending order of value. The lowest three tenders will be informed that their tenders are under consideration, and it is anticipated that an award will be made within 28 days of the opening of the tenders.

10. Tenders should be submitted to these offices by 10.00 a.m. on Tuesday 4 July 1992, and any tenders received after this time will not be considered. Tenders may be sent by post or delivered by hand, and must be in a plain envelope free from any identification of the name of the tenderer.

11. The Site may be visited during working hours on any day during the tender period. Appointments for visiting the Site should be made with Mr Brunel.

12. The cores from the Site survey are available for inspection at the Site.

The Institution of Civil Engineers
Conditions of Contract for Minor Works

Appendix to the Conditions of Contract

(to be prepared before tenders are invited and to be included with the documents supplied to prospective tenderers)

1. Short description of the work to be carried out under the Contract

   ....EXTENSION....TO....500 mm dia....MAIN....SEWER
   ....ROADS....AND....DRAINAGE....FOR....GALLOWS
   ....INDUSTRIAL....ESTATE..

2. The payment to be made under Article 2 of the Agreement in accordance with Clause 7 will be ascertained on the following basis. (The alternatives not being used are to be deleted. Two or more bases for payment may be used on one Contract.)

   ~~(a) Lump sum~~
   (b) Measure and value using a priced Bill of Quantities
   ~~(c) Valuation based on a Schedule of Rates (with an indication in the Schedule of the approximate quantities of major items)~~
   ~~(d) Valuation based on a Daywork Schedule~~
   ~~(e) Cost plus fee (the cost is to be specifically defined in the Contract and will exclude off-site overheads and profit)~~

3. Where a Bill of Quantities or a Schedule of Rates is provided the method of measurement used is

   ..CIVIL....ENGINEERING....STANDARD....METHOD....OF..
   MEASUREMENT 2nd EDITION (CESMM2)

4. Name of the Engineer
   (Clause 2.1) ..................GUY....COTTAM........................

5. Starting date (if known)
   (Clause 4.1) ....1st AUGUST 1990..................

6. Period for completion
   (Clause 4.2) ....12....WEEKS..............................

7. Period for completion of parts of the Works (if applicable) and details of the work to be carried out within each such part
   (Clause 4.2)

	Details of work	Period for completion
Part A	MAIN SEWER EXTENSION	..........6....WEEKS......
Part B	....	....
Part C	....	....

8. Liquidated damages
   (Clause 4.6) ....£ 500 PER WEEK..............

9. Limit of liquidated damages
   (Clause 4.6) ....£ 6000..............................

10. Defects Correction Period
    (Clause 5.1) ....12 MONTHS......................

11. Rate of retention
    (Clause 7.3) ....5 %.....................................

12. Limit of retention
    (Clause 7.3) ....£ 3000..............................

13. Minimum amount of interim certificate
    (Clause 7.3) ....£ 6000...................

14. Bank whose base lending rate is to be used
    (Clause 7.8) ..........BARCLAYS.....BANK...plc..............

15. Insurance of the Works
    (Clause 10.1) Required/~~Not required~~

16. Minimum amount of third party insurance (persons and property)
    (Clause 10.6) ....£..1...MILLION........................
    Any one accident/Number of accidents unlimited

Bill of Quantities

Daywork Schedule

Number	Item description	Unit	Quantity	Rate	Amount: £ p
	Method 1 - FCEC Schedule Rates				
1	**Daywork**				
	To be paid for at the rates set out in the Schedule of Dayworks carried out incidental to Contract Works issued by the Federation of Civil Engineering Contractors	Prov sum			25000

Number	Item description	Unit	Quantity	Rate	Amount: £ p
	Method 2 - FCEC Schedule Rates with Contractor's percentages (as CESMM)				
2	**Daywork**				
A411	Labour	Sum			4000
A412	Percentage adjustment to provisional sum for daywork labour	%			
A413	Materials	Sum			10000
A414	Percentage adjustment to provisional sum for daywork materials	%			
A415	Plant	Sum			6000
A416	Percentage adjustment to provisional sum for daywork plant	Sum			
A417	Supplementary charges	Sum			2000
A418	Percentage adjustment to provisional sum for daywork supplementary charges	%			

Number	Item description	Unit	Quantity	Rate	Amount: £ p
	Method 3 - A Contract Daywork Schedule				
3	Labour				
D1	General Operative	h			
D2	Carpenter	h			
D3	Steel Fixer	h			
D4	Plant Driver	h			
D5	Lorry Driver	h			
	Materials				
D6	Concrete grade 20	m^3			
D7	Softwood	m^3			
D8	Plywood B/BB quality 12 mm	m^2			
	Reinforcing steel, cut and bent				
D9	Mild steel to BS 4449 6 mm	kg			
D10	10 mm	kg			
D11	12 mm	kg			
D12	High yield to BS 4449 10 mm	kg			
D13	12 mm	kg			
D14	16 mm	kg			
	Plant				
D15	Excavator crane tracked full circle, slew up to 18 tonnes	h			
D16	Concrete tipping skip 20 m^3	h			
D17	Bar bending machine, hand operated	Day			
D18	Bar shearing machine, hand operated	Day			
D19	Lorry plated gross vehicle weight 7.6-12 t	h			

Appendix E

Preamble to CESMM 2

Amendments to standard methods of measurement

A. Amendments to the Civil Engineering Standard Method of Measurement, CESMM 2

When CESMM 2 is used in conjunction with the ICE minor works form it is necessary to make certain provisions in the preamble so that the two documents become compatible. A suggested outline for this Preamble is as follows

Preamble

1.0 Section 1. Definitions

In this document, and when construing the Civil Engineering Standard Method of Measurement (CESMM 2) in relation to it the following definitions shall apply.

1.1. 'Conditions of Contract' means the *Conditions of Contract, Agreement and Contract Schedule for use in connection with Minor Works of Civil Engineering Construction. First edition (January 1988)* commonly referred to as the ICE Conditions of Contract for Minor Works.
1.2. 'Provisional Sum' means a sum included in the Bill of Quantities for contingencies, which may be used in whole, or in part, or not at all at the instruction and discretion of the Engineer. The Contractor will be deemed not to have made any allowance in programming, planning or pricing his overhead costs for such work.
1.3. Since there is no provision for nominated sub-contractors in the Contract neither paragraph 5.5, nor Class A items 5** and 6** shall apply.
1.4. The words 'rate(s)' and 'price(s)' when used in the method of measurement shall both be included in the word 'prices' for the purposes of Clause 6.2.

2.0 Section 5. Preparation of the Bills of Quantities

2.1. The definition of rock in excavations is ....
2.2. Bodies of open water affecting the Works are:
(*a*) ....
(*b*) ....
(*c*) ....

3.0 Section 6. Completion pricing and use of the Bills of Quantities

3.1. Delete paragraph 6.4 and 6.5 and substitute new 6.4
3.2. For the purposes of Clause 7.3 interim additions or deduction on account of the amount, if any, of the adjustment item shall be made by instalments in interim certificates in the proportion that the amount that the Engineer considers due to the Contractor on the basis of the

monthly statement submitted under clause 7.2 before consideration of the adjustment item and after deducting any sums for materials on site bears to the total of the Bill of Quantities before the addition or deduction of the amount of the adjustment item.

4.0 Section 7. Method-Related Charges

Option A. For use when Method-Related Charges are omitted

4.1. Section 7 is omitted. The Contractor is not permitted to add items to the Bills of Quantities. Any attempt to do so may result in the rejection of the tender.
4.2. Items A3** of Class A General items do not apply.
4.3. Where any rule in the Work Classification provides for costs to be allocated to Method-Related Charges these costs shall be included in the rates and prices for the work for which they are needed.

Option B. For use when Method-Related Charges are included

4.1. Delete 7.6 and substitute new 7.6:
  7.6 Method-Related Charges shall not be subject to admeasurement but shall be deemed to be prices for the purposes of Clauses 6.1 and 6.2.
4.2. Delete 7.7 and substitute new 7.7:
  7.7 Method-Related Charges shall be certified and paid pursuant to Clauses 6.1 and 7.3.
4.3. Paragraph 7.8 Delete 'Clause 52' in the last line and substitute Clauses 6.1 and 6.2.

5.0 Specific amendments

Optional item. For use if General items are to be omitted.

5.1. Class A, General items is omitted. The rates and prices stated by the Contractor in the Priced Bill of Quantities shall (unless the Contract provides otherwise) cover all his obligations under the Contract.
5.2. Where any rule in the Work Classification provides for costs to be allocated to any item in Class A, General items these costs shall be included in the other items for the associated work.

B. Other methods of measurement

The major provisions that need to be examined when using any other method of measurement are those in relation to general items, nominated sub-contractors, provisional items, provisional sums and prime cost sums.

The Contract Schedule

(List of documents forming part of the Contract)

The Agreement (if any)

The Contractor's Tender (excluding any general or printed terms contained or referred to therein unless expressly agreed in writing to be incorporated in the Contract)

The Conditions of Contract

The Appendix to the Conditions of Contract

The Drawings. Reference numbers ....GE 90/101A, 102B......

...103A, 104, 105C, 106..

....

The Specification. Reference ....GC/ER.....PAGES...1-50.....

The priced Bill of Quantities*

~~The Schedule of Rates~~*

~~The Daywork Schedules~~*

The following letters*

from .... to .... dated ....

from .... to .... dated ....

from .... to .... dated ....

from .... to .... dated ....

*Delete if not applicable

Form of Tender

Short description of Works

All Permanent and Temporary Works in connection with[1] . . . .
. . . .
. . . .

Form of tender

To . . . .
. . . .
. . . .

Gentlemen

Having examined the Contract Schedule and all the documents listed therein for the construction of the above-mentioned Works we offer to perform and complete the whole of the said Works in conformity with the said Contract Schedule and documents

- (in accordance with 2(a) of the Appendix for a LUMP SUM contract) for the sum of . . . . (**state sum**)
- (in accordance with 2(b), (c), (d) or (e) of the Appendix) for such sum as may be ascertained thereunder.

We undertake to complete and deliver the whole of the Permanent Works comprised in the Contract within (**the time stated in the Appendix to the Conditions of Contract**) (. . . . . **weeks from the Starting Date**)

If our tender is accepted

- We propose to place the Insurance for Public Liability in accordance with Clause 10.6 with (**name of insurance company**)
- We have in force an annual insurance policy covering [**care of the Works to their full value with the** (**name of insurance company**)] [**and**] [**a policy for our Public Liability to the value of £. . . . . .** (**state sum**)] for any one accident, and the number of accidents is unlimited, with (**name of insurance company**).

We undertake to sign the Agreement when requested to do so, and until it is executed this tender together with your written acceptance thereof shall constitute a binding contract between us.

Our tender is open for acceptance for 56 days. We understand that you are not bound to accept the lowest or any tender you may receive.

We are, Gentlemen,

Yours faithfully,

(**Signature**)
(**Address**)
(**Date**)

---

1. Complete or delete as appropriate.

2. Letter of acceptance

HITECH & PARTNERS
Consulting Engineers
37–39 High Street, East Cheam
Avon, BS27 8QD

(Name of Contractor)
(Address of Contractor)

(Date)

Dear Sir/Madam

Re: (Title of Contract)

(I/We) have been instructed by **(name of Employer)** to accept your tender dated **(date of tender)** for these Works.

Enclosed with this letter are two copies of the Agreement signed by the Employer. Please sign and date both copies and return one copy to **(me/us)**. The other copy is for your retention.

Please advise **(me/us)** of the name of the insurance company and the wording of the policy you propose for the public liability insurance, required in accordance with Clause 10.6, for the Employer's approval.

Please forward copies of your insurance policies required by Clauses 10.1 and 10.6 of the Contract for the Employer's approval.

The Starting Date for the Works will be **(Starting Date)**.

Yours faithfully

(Name of Engineer)

3. FCEC Form of Sub-Contract

The parties

AN AGREEMENT made the ....FIRST............ day of ....JANUARY........ 19 ..90......

BETWEEN ..........MAINCON.....CONSTRUCTION.....Ltd........................... of/whose

registered office is at ...MAINCON...HOUSE,....FREEWAY,......LONDON....N14......

.... (hereinafter called "the Contractor") of the one

part and ....MACTAR.....Ltd.. of/whose

registered office is at ....23....HIGH....ROAD,.....BYWAY,....ESSEX.............................

.... (hereinafter called "the Sub-Contractor") of the other part.

WHEREAS the Contractor has entered into a Contract (hereinafter called "the Main Contract") particulars of which are set out in the First Schedule hereto:

AND WHEREAS the Sub-Contractor having been afforded the opportunity to read and note the provisions of the Main Contract (other than details of the Contractor's prices thereunder), has agreed to execute upon the terms hereinafter appearing the works which are described in the documents specified in the Second Schedule hereto and which form part of the works to be executed by the Contractor under the Main Contract:

NOW IT IS HEREBY AGREED as follows:

Definitions.

1. (1) In this Sub-Contract (as hereinafter defined) all words and expressions have the same meaning as in the Main Contract unless otherwise provided or where the context otherwise requires.

(a) "the Main Contract" means the contract, particulars of which are given in the First Schedule hereto.

(b) "the Sub-Contract" means this Agreement together with such other documents as are specified in the Second Schedule hereto, but excluding any standard printed conditions that may be included in such other documents unless separately specified in the said Schedule.

(c) "the Sub-Contract Works" means the works described in the documents specified in the Second Schedule hereto.

(d) "the Main Works" means the Works as defined in the Main Contract.

(e) "the Price" means the sum specified in the Third Schedule hereto as payable to the Sub-Contractor for the Sub-Contract Works or such other sum as may become payable under the Sub-Contract.

(2) Words importing the singular also include the plural and vice-versa where the context requires.

(3) The headings, marginal notes and notes for guidance in this Agreement shall not be deemed to be part thereof or taken into consideration in the interpretation or construction thereof or of the Sub-Contract.

(4) All references herein to clauses are references to clauses numbered in this Agreement and not to those in any other document forming part of the Sub-Contract unless otherwise stated.

General.

2. (1) The Sub-Contractor shall execute, complete and maintain the Sub-Contract Works in accordance with the Sub-Contract and to the reasonable satisfaction of the Contractor and of the Engineer.

(2) The Sub-Contractor shall provide all labour, materials, Constructional Plant, Temporary Works and everything whether of a permanent or temporary nature required for the execution, completion and maintenance of the Sub-Contract Works, except as otherwise agreed in accordance with Clause 4 and as set out in the Fourth Schedule hereto.

(3) The Sub-Contractor shall not assign the whole or any part of the benefit of this Sub-Contract nor shall he sub-let the whole or any part of the Sub-Contract Works without the previous written consent of the Contractor.

Provided always that the Sub-Contractor may without such consent assign either absolutely or by way of charge any sum which is or may become due and payable to him under this Sub-Contract.

Main Contract.

3. (1) The Sub-Contractor shall be deemed to have full knowledge of the provisions of the Main Contract (other than the details of the Contractor's prices thereunder as stated in the bills of quantities or schedules of rates and prices as the case may be), and the Contractor shall, if so requested by the Sub-Contractor, provide the Sub-Contractor with a true copy of the Main Contract (less such details), at the Sub-Contractor's expense. The Main Contractor shall on request provide the Sub-Contractor with a copy of the Appendix to the Form of Tender to the Main Contract together with details of any contract conditions which apply to the Main Contract which differ from the standard ICE Conditions of Contract.

(2) Save where the provisions of the Sub-Contract otherwise require, the Sub-Contractor shall so execute, complete and maintain the Sub-Contract Works that no act or omission of his in relation thereto shall constitute, cause or contribute to any breach by the Contractor of any of his obligations under the Main Contract and the Sub-Contractor shall, save as aforesaid, assume and perform hereunder all the obligations and liabilities of the Contractor under the Main Contract in relation to the Sub-Contract Works.

For a sealed contract

(8) Without prejudice to any other remedies available to either the Contractor or the Sub-Contractor the Contractor shall be entitled to withhold from the Sub-Contractor an amount of money equal to the amount of any tax paid to the Sub-Contractor in excess of the amount which in all the circumstances should have been paid to the Sub-Contractor.

(9) Where the Contractor makes any provision pursuant to Clause 4 (Contractors' Facilities) and that provision is a taxable supply to the Sub-Contractor by the Contractor the Sub-Contractor shall pay tax to the Contractor in accordance with the requirements of the Contractor at the rate properly payable by the Contractor to the Commissioners.

Law of Sub-Contract.

20. The Law of the Country applying to the Main Contract shall apply to this Sub-Contract.

*IN WITNESS whereof the parties hereto *have caused their respective Common Seals to be hereunto affixed*

~~*have hereunto set their respective hands and Seals*~~.

the day and year first above written.

* *Strike out the words not applicable and use appropriate execution clause on page 16. If the Sub-Contract is executed under Seal a fifty pence stamp must be affixed within 30 days of the date of the Sub-Contract.*

FIRST SCHEDULE

PARTICULARS OF MAIN CONTRACT

Parties: THE EMPLOYER: LONDON BOROUGH OF HEAVE HO
THE CONTRACTOR: MAINCON CONSTRUCTION Ltd.
THE CONTRACT INCORPORATES THE FOLLOWING DOCUMENTS:
THE CONTRACTOR'S TENDER DATED 1st. MARCH 1989
THE EMPLOYER'S LETTER OF ACCEPTANCE DATED 30th. JULY 1989
GENERAL CONDITIONS OF CONTRACT [AS AMENDED]
SPECIFICATION [REF: CE WORKS 11/80]
DRAWINGS [AS LISTED IN APPENDIX A]

Date: BILLS OF QUANTITIES 1 - 8

Brief Description of Main Works:

THE WORKS INVOLVE DUALLING 3.2 km OF THE A33491 BETWEEN THE ROUNDABOUT AT GALLOWS CORNER AND THE HEAVE HO BYPASS. THE WORK ENTAILS IMPROVEMENT OF THE EXISTING ROUNDABOUT, DRAINAGE, AND ANCILLARY WORKS. THE EXCAVATION OF ABOUT 60,000 M3 OF SUITABLE MATERIAL OF WHICH 30,000 M3 WILL BE USED AS FILL. PART REPLACEMENT OF TRAFFIC SIGNAL AND STREET LIGHTING EQUIPMENT.

Date(s) on which Main Contractor will be submitting statements to the Employer — "Specified Date".

LAST FRIDAY OF EACH MONTH

Minimum amount of Interim Certificates under Main Contract:

£15,000

SECOND SCHEDULE

(A) Further Documents Forming part of the Sub-Contract

1. SPECIAL CONDITIONS Nr. 1 - X.
2. SPECIFICATION PAGES 63 - 108.
3. BILLS OF QUANTITY PAGES 35 - 42 AS PRICED BY THE SUB-CONTRACTOR WITH HIS TENDER DATED 1.12.89
4. DRAWINGS Nr. 1109/37, 39, 41A + 43B

(B) Sub-Contract Works

SUPPLY AND LAY ROAD BASE, FLEXIBLE SURFACING AND WEARING COARSE, FOOTPATH SURFACING, SCARIFYING, PLANING AND BURNING OFF WORK.

(C) Fluctuation Provisions (if any) affecting payments under this Agreement

THE PRICE IS FIXED FOR THE DURATION OF THE CONTRACT

THIRD SCHEDULE

(A) ~~The Price~~ .... /Measure and Value

(B) (i) Percentage of Retention — Works

5%

(ii) Percentage of Retention — Materials on Site

10%

(iii) Limit of Retention

£3,000

(C) Period for Completion

14 WEEKS
THE SUB-CONTRACTOR HAS INCLUDED FOR THREE VISITS. ADDITIONAL VISITS WILL INCUR A MOBILIZATION COST OF £500.

THE SUB-CONTRACTOR WILL NOT BE REQUIRED TO RE-VISIT THE SITE IF THERE IS LESS THAN THREE DAYS' WORK AVAILABLE.

FOURTH SCHEDULE

Contractor's Facilities

PART I — Common Facilities (Clause 4(2))

(A) Constructional Plant	Terms and Conditions
5 TONNE MAX CAPACITY TOWER CRANE	WHEN AVAILABLE MINIMUM OF 2 HOURS NOTICE SHOULD BE GIVEN WHEN USE IS REQUIRED. CHARGE £25 PER HOUR. HOURS TO BE AGREED WITH THE DRIVER AND RECORDED IN WRITING AT THE CONCLUSION OF EACH USE
(B) Other Facilities	
ELECTRICITY SUPPLY	50 VOLTS MAXIMUM CAPACITY 50 AMPS UNITS FREE BUT SUB-CONTRACTOR MUST ENDEAVOUR TO MINIMIZE UNITS USED.
WATER	25mm SUPPLY WILL BE PROVIDED CENTRALLY. SUB-CONTRACTOR TO PROVIDE ANY REQUIRED DISTRIBUTION.
TOILET FACILITIES	THE CONTRACTOR WILL PROVIDE ADEQUATE TOILET FACILITIES FOR ALL PERSONS ON THE SITE. WASHING FACILITIES WILL BE LIMITED TO HAND BASINS.

Contractor's Facilities

PART II — Exclusive Facilities (Clause 4(3))

(A) Constructional Plant	Terms and Conditions
250 cfm COMPRESSOR	£15 per HOUR
2" CENTRIFUGAL PUMP	£10 PER HOUR
TRAFFIC LIGHTS	£20 PER DAY
	ALL PLANT IS SUBJECT TO AVAILABILITY, AND NO GUARANTEE IS GIVEN THAT IT WILL BE AVAILABLE AT ALL.
(B) Other Facilities	
STORAGE COMPOUND	THE CONTRACTOR WILL PROVIDE AN AREA 20m X 30m. THE SUB-CONTRACTOR IS RESPONSIBLE FOR ALL ACCOMMODATION, FENCING, GATES AND SURFACING etc THAT HE REQUIRES.

FIFTH SCHEDULE

INSURANCES

Part I Sub-Contractor's Insurances

Third Party Liability - £500 000 (Five Hundred Thousand Pounds).

Employer's Liability - Unlimited.

Insurance against loss or damage of the Sub-contractor's plant and equipment (whether owned or not) and all the Sub-contractor's property and personal effects of his Employees.

Part II Contractor's Policy of Insurance

CONTRACTOR'S ALL RISKS INSURANCE

SUM INSURED:	subject to an excess of: (a) £1000 each and every loss (b) £ 250 each and every loss in resect of Temporary Buildings, Office Contents, non-mechanical plant and scaffolding. (c) £ 10 each and every loss in respect of Employees' Personal Effects and Tools.
IN RESPECT OF:	Gallows Corner and all ancillary works. On Contract Works, Temporary Works, Materials, Machinery and all other things brought into the Contract area for the purposes of the Contract. Including: (1) Inland Transits and Storage Risks anywhere in the United Kingdom. (2) Architects,/Engineers/Consultants/professional fees. (3) Removal of Debris (in addition to sum insured limit). (4) Temporary Buildings and Office Contents, non-mechanical plant and scaffolding whether owned or hired-in being used for contracts, and whilst in storage at Insurer's premises. (5) Personal effects for limits of £250 in respect of tools and £200 effects per person subject to a maximum of £2000 on any one site. Excluding Contractor's mechanical plant and equipment.

THE SUB-CONTRACTOR WILL HAVE NO RIGHT TO INDEMNITY UNDER THIS POLICY FOR LOSS OR DAMAGE TO THE SUB-CONTRACTOR'S PLANT AND EQUIPMENT (WHETHER OWNED OR NOT) OR TO THE SUB-CONTRACTOR'S PROPERTY, OR TO THE PERSONAL EFFECTS OF HIS EMPLOYEES.

Sealed contract

THE COMMON SEAL of the above-named Contractor was hereunto affixed in the presence of:—

L.S.

....

....

THE COMMON SEAL of the above-named Sub-Contractor was hereunto affixed in the presence of:—

L.S.

....

....

OR

SIGNED SEALED AND DELIVERED AS A DEED by the above-named Contractor in the presence of:—

....

....

SIGNED SEALED AND DELIVERED AS A DEED by the above-named Sub-Contractor in the presence of:—

....

....